European Foundation
for the Improvement of
Living and Working Conditions

From Drawing Board to Building Site

WORKING CONDITIONS
QUALITY
ECONOMIC PERFORMANCE

London: HMSO

Luxembourg: Office for Official Publications of the European Communities

This booklet is the synthesis of six national studies which were carried out on behalf of the European Foundation for the Improvement of Living and Working Conditions.

1. The Management College, D W Birchall, *Working in the Construction Industry*, Henley, March 1987.

2. CERTES, E Campagnac and C Caro, *Les conditions de travail dans l'industrie de la construction*, Paris, March 1987.

3. Stichting Bouwresearch, J Diepeveen and A Brouwers, *Les conditions de travail dans l'industrie de la construction*, Rotterdam, March 1987.

4. Comité National d'Action pour la Sécurité et l'Hygiène dans la Construction, P. Lorent, *Les conditions de travail dans l'industrie de la construction – Productivité, conditions de travail, qualité concertée et totale*, Brussels, April 1987.

5. RSO, M Rollier, *La sicurezza nel settore delle costruzioni come problema di organizzazione e di progettazione*, Milan, February 1987.

6. Projektgruppe Arbeitsbedingungen in der Bauwirtschaft (PAB), B Spannhake, *Les carences de la sécurité du travail dans l'industrie de la construction et les frais qui en découlent*, Dortmund, March 1987.

This document was drawn up by Pierre Lorent of the CNAC on the basis of the observations of a working group made up as follows: Elisabeth Campagnac, David Birchall, Jos Kooren, Matteo Rollier, Brunnhilde Spannhake and co-ordinated by Pascal Paoli. Technical illustrations: P Gheux. Cartoons: D. Billout.

Publication no. EF/88/17/FR of the European Foundation for the Improvement of Living and Working Conditions, Loughlinstown House, Shankill, Co. Dublin, Ireland

First published in English 1991.

HMSO, London

ISBN 0 11 701576 8

Office for Official Publications of the European Communities, L-2985 Luxembourg

ISBN 92-825-8685-5

Catalogue number SY-53-88-027-EN-C

Printed in the United Kingdom for HMSO
Dd291782 10/91 C23 G3392 10170

CONTENTS

PREFACE

At a time when the completion of the single market makes it essential to reinforce economic and social cohesion, it is important to emphasise that working conditions constitute not only a fundamental issue, but also an essential condition of competitiveness. Far from being merely a matter of costs arising in advance or in arrears, the improvement of working and safety conditions should be seen as a resource to be exploited for productivity and enhanced performance.

This means that economic and social imperatives are not necessarily contradictory. On the contrary, they are convergent, and the present work brings out that convergence and the contribution they can make in the construction sector.

In fact, the far-reaching upheavals affecting this sector, which are being accentuated by the opening up of frontiers, demand a more detailed mastery over timescales and costs, particularly the quality of the product and the way in which it matches the client's requirements.

In this regard, fundamental importance is attached to the need for better consideration of the conditions under which work is carried out on the site, and better forward management of the workforce in terms of jobs, qualifications and work organisation. In this area, a certain number of paths are open to consideration by contracting authorities, project managers, company chiefs, trade union leaders and training personnel. Beyond all this, the European Foundation for the Improvement of Living and Working Conditions hopes to make a contribution to the area of negotiation and social dialogue.

Clive Purkiss
Director

Erich Verborgh
Deputy Director

INTRODUCTION

The construction sector, especially building, is a sector undergoing far-reaching changes:

Technical change, with the emergence and application of new construction procedures, new equipment and materials.
Market change, with a reduction in the number of large-scale housing programmes, and a change away from the construction of new buildings and towards renovation and restoration in most Community countries.
Organisational change, with a search for productivity gains no longer confined to the level of each separate sequence featuring in the construction project, but also looking at the overall cohesion of the process itself.

This overall approach relies, in particular, on improved communication between the different participants, both upstream and downstream of the process.
Structural change, finally, with a trend developing in some countries towards the breakdown of functions and the fragmentation of companies. This leads to a cluster of subcontracting arrangements around a supervising employer, or general contractor, pared down to its most basic form.

These changes, which other sectors of industry have experienced before the construction sector, require preparation and accompanying action. This must be done in a dual perspective:

The first aspect is connected with **the opening of frontiers and the creation of the single internal market in 1992**. Ability to control the technical studies, the planning and organisation of the works, the management of site work, the overall co-ordination and hence the control of costs and quality, is going to lead to a new division of roles within the Community.
The second aspect is connected with **the social dimension and the specific problems experienced by the sector in this area**.

It is not an overstatement to say that – in spite of improvements – the sector is still characterised by unpleasant working conditions and a higher degree of gravity and frequency of occupational accidents than in other industrial sectors. The resulting social and economic costs are extremely high.

Increased reliance on subcontractors and self-employed people, leading to a dilution of responsibilities for training and accident prevention, undoubtedly undermines the major efforts made in recent years in the area of regulation and monitoring. That is why, without relaxing the efforts in this area, it is necessary to find other ways of taking action, especially upstream of the construction site in the strict sense.

In this context, a unique opportunity exists at the present time to see that this social dimension (working conditions, organisation and execution of work, skills required) is not simply an adjustable variable, but rather one of the motors of change and one of the tools favouring the emergence of new modes of productive organisation, leading to both social and economic progress.

The aim of this book is to illustrate, through examples, the close relationship between the conditions under which work is done on the construction site, and performance, i.e. the quality of the finished product (its conformity with the customer's requirements) and the control of costs, including the time involved.

It is a commonplace to say that the site is the point of convergence for all malfunctions created upstream, and that this is where the price is paid for delays, errors and omissions in the study and planning phases, with all the well-known effects on companies (time constraints, difficulties with implementation, etc.). Now, each function, from the beginning to the end of the project – in other words, from the definition of needs to the completion of construction – is affected by cost control and quality improvement.

The identification of hidden costs and their sources is of fundamental importance, and it is clear that working conditions are at the heart of the new organisation of production in the construction industry. The organisation of site work, the skills envisaged and the prevention of accident risks thus become instruments for optimisation and necessary stages in the modernisation of the sector. In this perspective, the social parameters become a resource available to all of the participants in accomplishing their mission, without this relieving the companies of their responsibilites for prevention in the areas of safety and health.

CHAPTER 1
THE ISSUES AT STAKE

1.1 CHANGES GIVING RISE TO NEW REQUIREMENTS

1.1.1 DEVELOPMENT OF THE MARKET

1.1.1.1 Towards a revival of activity?

Some Community Member States are witnessing a slight economic upturn in the construction sector at the present time. This is connected with a growth in demand on the part of the private sector, together with a developing market for restoration and renovation, accompanied by the need for a more highly qualified workforce.

1.1.1.2 Orientation of the housing market towards renovation all over Europe

In fact, the price of existing houses (the secondary market) generally stands at a lower level than the price of new houses (the primary market).

1.1.1.3 Buyer's market/seller's market

The construction market has undergone a fundamental structural change, from a seller's to a buyer's market, essentially on account of the reduction in big public construction programmes. In addition, formerly all surges in the economy took place in conjunction with the construction sector, or under pressure from it. This is no longer the case today.

1.1.2 END OF LONG PRODUCTION RUNS AND REPETITIVE BUILDING SITES

Accompanying the end of long production runs and repetitive building sites, the constraints posed by clients are becoming at once more varied and more weighty, whether in the areas of environment, ergonomics, user safety, energy performance, etc. This forces builders to be more flexible in responding to the multiplicity of needs put forward by investors.

1.1.3 OPENNESS OF THE NEW SINGLE EUROPEAN MARKET

The completion of the single market will accelerate the opening out of the sector. Important issues emerge in this setting, both on the technical and the organisational side, and there is a possibility that a new division of roles may emerge. In fact, it will no longer be at all rare to find, for example, a Danish architect working with a French company, using Italian subcontractors, who themselves use German equipment.

1.1.4 CONCLUSIONS

These new changes demand that the sector should show more flexibility and innovation to respond to a more varied European market where the culture, the social climate, the training of workers, local products, etc. determine the mode of design, organisation and implementation on site.

The competitive environment in this new Europe will force companies, main contractors and also clients to bring together their skills to improve the productivity of building operations by integrating, from the earliest phases of project orientation, parameters which will make it possible to secure enhanced control over the relationship between costs and desired quality, from beginning to end of the project.

1.2 STRUCTURES OFTEN UNSUITED TO THESE NEW REQUIREMENTS

1.2.1 INCREASING EMPHASIS ON SPECIALISATION AND STRUCTURAL FRAGMENTATION

1.2.1.1 At the contract stage

Companies with major financial resources are acting the role more and more of 'general contractor' or overall enterprise:

– by co-ordinating the network of producers and service providers;
– by confining themselves to controlling lead times and the match between the finished product and the requirements formulated by the client (5)*.

1.2.1.2 At the site stage

The work is overwhelmingly given to small teams, sometimes organised into family-based mini-businesses, using equipment provided by the contracting company.

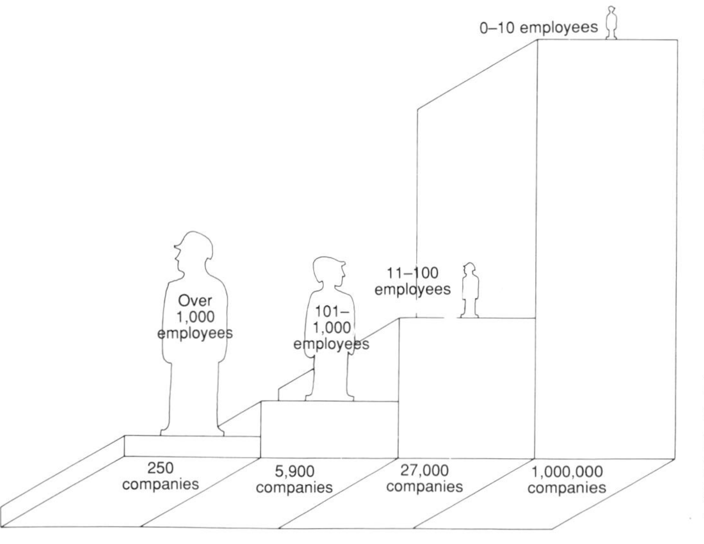

Fig. 1.1 Number of companies: breakdown by size.
Out of the 1,102,150 companies surveyed in the EEC, 250, or 0.03%, have more than 1,000 employees
5,900 or 0.53%, have between 101 and 1,000 employees
27,000, or 8.79%, have between 11 and 100 employees
1,000,000, or 90.65%, have between 0 and 10 employees (7)

* Numbers in parentheses refer, throughout the book, to entries in the Bibliography.

1.2.2 MULTIPLICITY OF SMALL FIRMS

The multiplicity of small firms and groups of subcontractors is causing a dilution of responsibility and a lack of co-ordinated policy (Fig. 1.1).

1.2.2.1 The qualification of construction workers

The average age of the workforce in the sector is showing a general tendency to increase:

- the mobility of workers aged over twenty-five has diminished sharply;
- young people are turning more to other sectors because of working and employment conditions;
- in most large companies, the average age of the workers is over thirty-five;
- the number of construction schools is declining sharply, and very soon the sector will have a shortage of competent young people, both on the management and worker sides.

1.2.2.2 The prevention of accidents at work

The construction industry:

- employs 7 per cent of all employees in the Community;
- accounts for 15 per cent of all occupational accidents;
- 30 per cent of all fatal accidents in the industrial sector (8).

1.2.2.3 Precariousness of employment

Bankruptcies in this sector are often signs of manifest errors of management. These are translated into high costs which are spread over the early and late stages of the project, both from a social and an economic point of view, within the companies and affecting the project as a whole.

1.3 HIGH SOCIAL AND ECONOMIC COSTS

1.3.1 THE COSTS OF POOR QUALITY

These represent 10–18 per cent of the works, affect all activities on site and are spread over the entire process (9).

1.3.2 COSTS OF OCCUPATIONAL ACCIDENTS

These represent 3 per cent of the turnover of companies, or between 7 and 10 per cent of the salary bill in the sector, if one includes the salary and rehabilitation costs as well as the indirect costs caused by an accident (10). But the average profitability of construction companies in Europe comes to 1.3 per cent of turnover (less than half of the costs caused by accidents). Only through an awareness of these shortcomings in product quality and working and safety conditions can the participants generate profitable procedures from beginning to end of the project.

1.4 CONCLUSIONS

The new requirements of the market connected with

- an increase in renovation activities;
- the ending of large-scale repetitive mass production;
- the introduction of innovative techniques in companies and suppliers of equipment and materials;
- new structures for participation on building sites;
- a demand from clients investing in construction which is more directed towards quality;
- lastly, the completion of the internal market in 1992;

mean that companies, contracting authorities and general contractors have to reflect on the future so that they can make the construction sector into a modern and competitive sector. There have been advances in productivity in the construction sector in recent years, but this has happened much more slowly than in other industrial sectors which have already undergone restructuring. This means that a new rationale must be devised, based on control over the whole process instead of isolated management of each sequence in the process (see Chapter 5).

The current development of new configurations in the design, organisation and implementation of construction projects, shows the emergence of a new model of industrial organisation, and leads to a redefinition of the key stages in the management of projects and the role of the different participants. It also leads to a better understanding and mastery of the implications of decisions made before the site work begins, as regards working conditions, productivity and the quality of the finished project.

CHAPTER 2
HIGH HIDDEN COSTS WHICH MUST BE IDENTIFIED

Besides the direct costs which can easily be identified, a certain number of costs do not appear directly, or else appear only partially, in connection with compatibility between companies. These hidden costs, an understanding of which is essential for improving productivity and work quality in the sector, are high (Fig. 2.1). Control of these costs is all the more essential given that profit margins are low, as we have already noted. These costs are also indicators of malfunctions at all levels of construction projects, as we shall see, and thereby concern all personnel associated with the project, from the main contractor to the subcontractors, and including the architects and consultants involved.

We will review three types of social and economic costs. These are connected with quality (or shortcomings in the quality level), in the broad sense, with occupational accidents and with working conditions on the site. There is obviously a close relationship between these various factors. The object of Chapter 2 is thus to show that there are considerable margins of possible gains, spread over the earlier and later stages of the project. Subsequently (Chapter 3), the functions and personnel concerned in controlling these different costs will be identified.

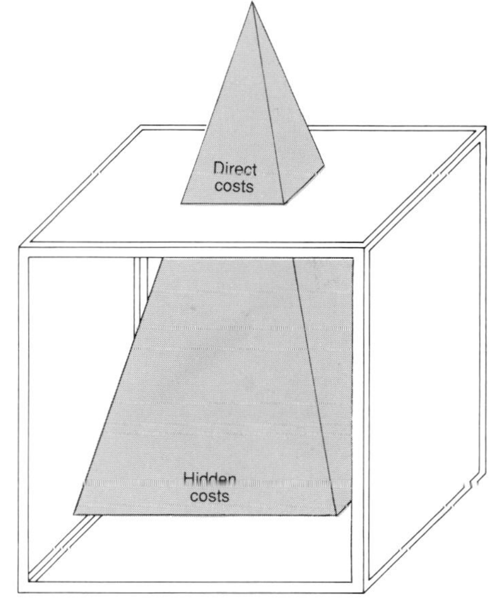

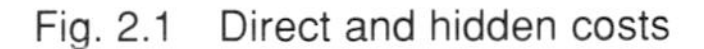
Fig. 2.1 Direct and hidden costs

2.1 COSTS OF SHORTCOMINGS FOUND ON SITE

2.1.1 INDICATORS

These have to do with the design, organisation and planning processes upstream of the building site, and with the implementation of the works. They are divided between the operations and participants as shown in Table 2.1.

Table 2.1

	The participants		
The operations	The client	The contractor	The companies
Cost of extra consultancy work Loss in implementation Repairs Unnecessary works Expert advice on accidents	Number of units unsold Cost of disputes Operational delays Badly defined programme Extra costs due to bad choice of partners Defective targeting of customers	Number of plans, meetings and disputes Errors in design and organisation	Days delayed Hours' repairs Equipment breakdowns Excessive consumption of materials Defective co-ordination Financial losses Extra quality Errors and malfunctions

2.1.2 COSTS

It is admitted that the cost of shortcomings can reach somewhere between 10 and 18 per cent of turnover, i.e. between 20 and 45 per cent of the salary bill in the sector (4).

2.1.3 CAUSES

Application to a five-level building (Fig. 2.2) (see example on following pages).

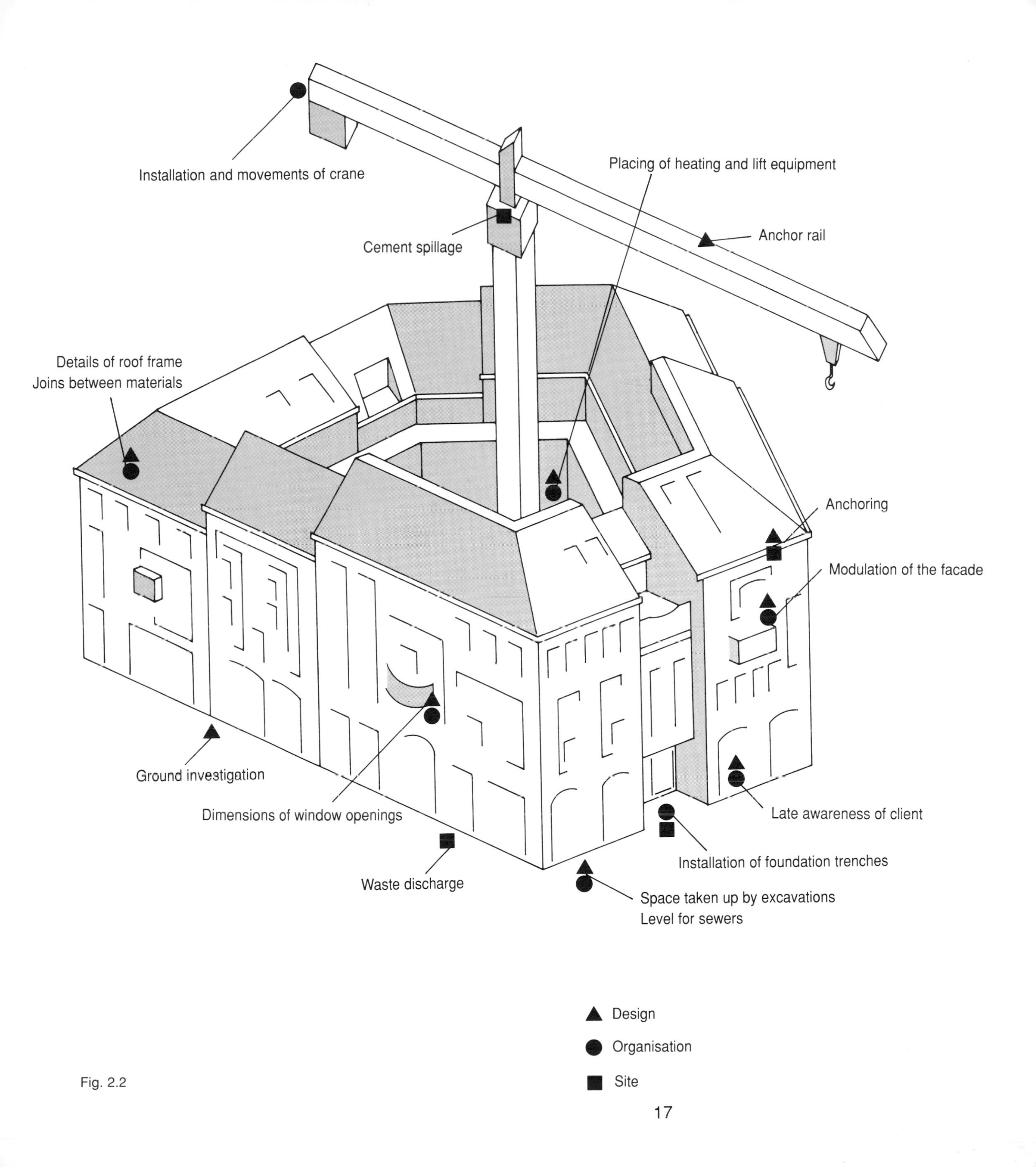
Installation and movements of crane
Placing of heating and lift equipment
Anchor rail
Cement spillage
Details of roof frame
Joins between materials
Anchoring
Modulation of the facade
Ground investigation
Dimensions of window openings
Late awareness of client
Installation of foundation trenches
Waste discharge
Space taken up by excavations
Level for sewers
Design
Organisation
Site

Fig. 2.2

EXAMPLE

CONSTRUCTION OF A BUILDING WITH COMMERCIAL GROUND-FLOOR LEVEL AND FIVE STOREYS OF APARTMENTS

COST OF FINISHED BUILDING: 529,000 ECUs

Without the shortcomings noted, this building could have been completed for the sum of 483,688 ECUs (a saving of 8.6% of the total cost of works). (See Table 2.2 and Fig. 2.3.)

We have classified the costs of shortcomings and their prevention under three headings:

– costs arising from design errors;
– costs arising from organisation errors;
– costs arising from implementation errors.

Table 2.2 Extra costs of poor quality and prevention costs

The cost of non-intervention upstream	The cost of intervention upstream
Extra costs of poor quality 11% of the total cost of works	*Extra preventive costs to carry out the works correctly* 2.4% of the total cost of works
Design phase: 8% Organisation phase: 2% Implementation phase: 1%	Design phase: 2% Organisation phase: negligible Implementation phase: 0.4%
'Downstream' costs: 11%	'Upstream' costs: 2.4%

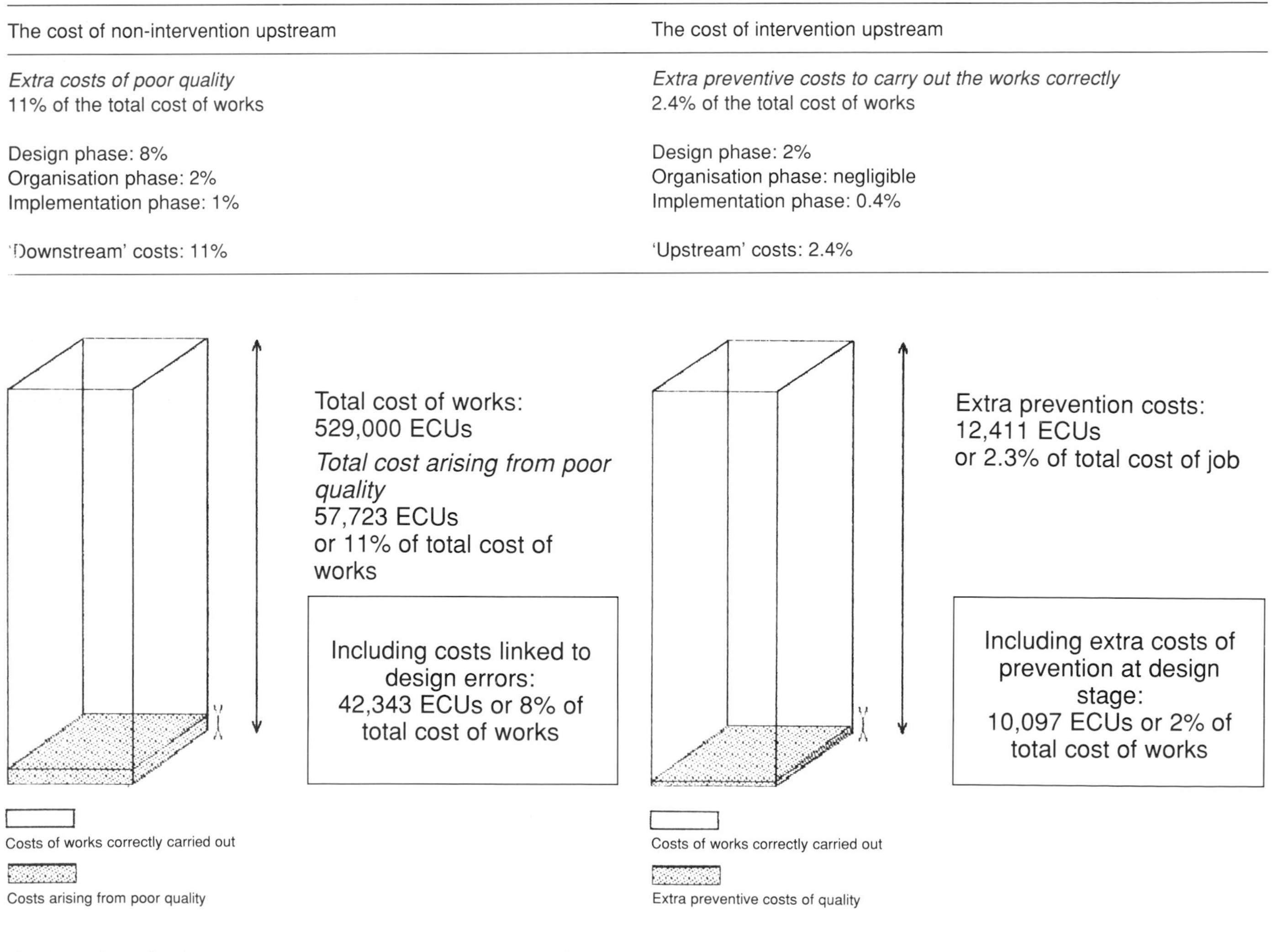

Fig. 2.3 Overall calculation of costs of poor quality and prevention costs for a building

DESIGN

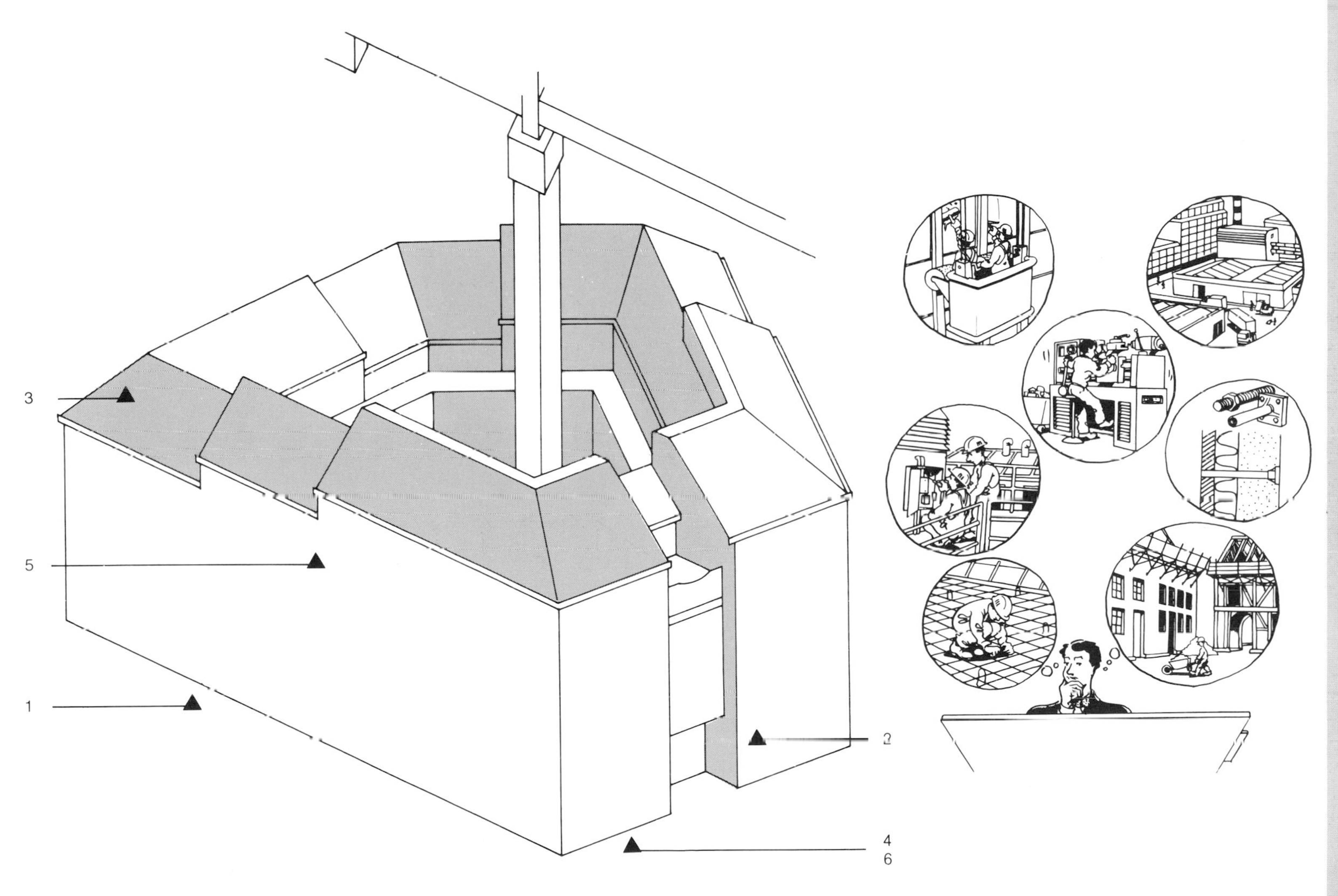

Fig. 2.4

PROJECT DESIGN – EQUIPMENT DESIGN (Fig. 2.4)

Costs arising from errors in design	Prevention costs
1. *Lack of ground investigation – modified foundations*	
It was not felt necessary to investigate the ground strata at the design stage. When excavations were made, the state of the ground dictated a new type of foundation. *Consequences:* – fresh surveys – different type of foundations 9,798 ECUs	Ground investigation carried out in good time 1,656 ECUs
2. *An unforeseen client*	
The commissioning client was unable to tell the architect exactly what type of commercial activities would take place on the ground floor The installation of a strong-room after the completion of the works caused considerable problems in launching the commercial ground floor. The concreting work proved particularly difficult *Consequences:* – disturbances due to works, failure to attract traders: not quantifiable – supplementary surveys, demolition of concrete curtain walls, cement work, tax for occupying public thoroughfare 28,750 ECUs	If the client had been known at the design stage: – supplementary fees for the architect – main building works 8,280 ECUs
3. *Worker protection: installation and maintenance of roofing*	
Carried out after the event, this protection would have cost 506 ECUs	To facilitate the work of roofers and asphalters, to comply with legislation on protection for roofing accessible on an exceptional basis, the architect designed anchor rails set into the cement of the top pediment. These rails allow a handrail to be installed, together with a groove for attaching a safety harness 161 ECUs

Costs arising from errors in design | Prevention costs

4. *More detailed study of the foundations*

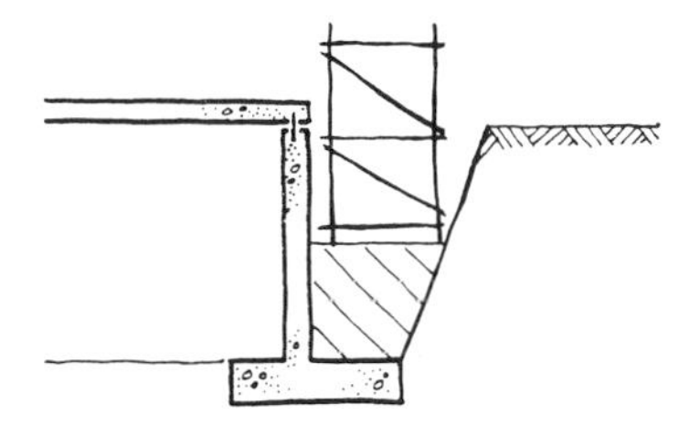

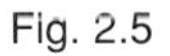

Fig. 2.5

Without this modification (Fig. 2.5), the works would have cost an extra 1,495 ECUs

To reduce the scale of the earthworks, the project director proposed a different type of foundation, facilitating access to the site, reducing the problem of scaffolding, and decreasing the site's obstruction of the public thoroughfare by three weeks (Fig. 2.6).

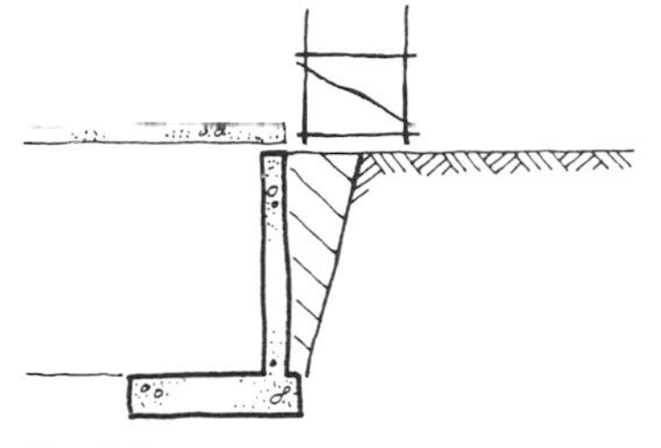

Fig. 2.6

5. *Modification of facades*

Social cost:

Inadequate protection for workers

Economic cost:

Use of temporary protection devices which do not offer protection when structures are being installed.
Unquantifiable gain (Fig. 2.7)

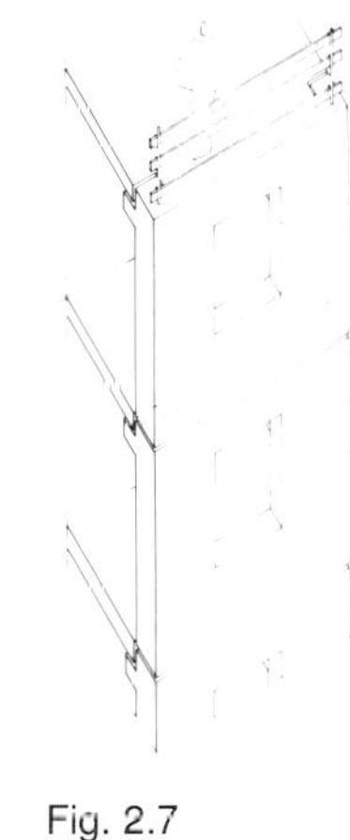

Fig. 2.7

The project designer modified the architectural cement facades so as to integrate the protection of workers against falls. This made it possible to avoid using temporary guard-rails which in any case would have hindered the positioning of the next level on top. Productivity and safety in the assembly phase were improved by this procedure (Fig. 2.8)

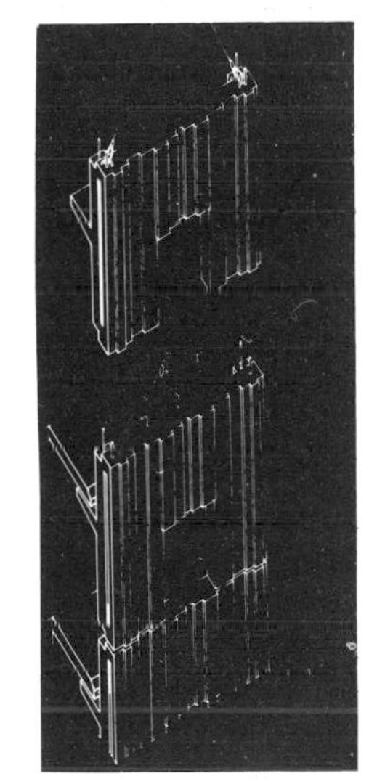

Fig. 2.8

6. *Mistakes in gauging levels*

The building is set up on levels specified by the architect, who, it was assumed, had checked the level of the sewers. When the foundations were completed, and drains were being installed, it transpired that the main sewers were too high.

Consequences:

Discussions with the architect, construction of a sump, water pump, new mains
1,794 ECUs

A check on the depth of the main sewer when earthworks were taking place would have allowed the compatibility of levels to be ascertained.
The levels of sewerage in the building could have been modified without cost
Nothing

OVERALL CALCULATION OF THE COSTS OF POOR QUALITY AND THE COSTS OF PREVENTION IN CONSTRUCTING A BUILDING

Total cost of works 529,000 ECUs

Total cost arising from poor quality 57,723 ECUs or 11% of the total cost of works	*Extra prevention costs* 12,411 ECUs or 2.4% of the total cost of works
Including costs arising from design errors 42,343 ECUs 8% of the total cost of works	*Including extra costs at design stage* 10,097 ECUs 2% of the total cost of works
Costs arising from organisation errors 10,028 ECUs 2% of the total cost of works	*Extra prevention costs at the organisation phase* 218 ECUs Negligible

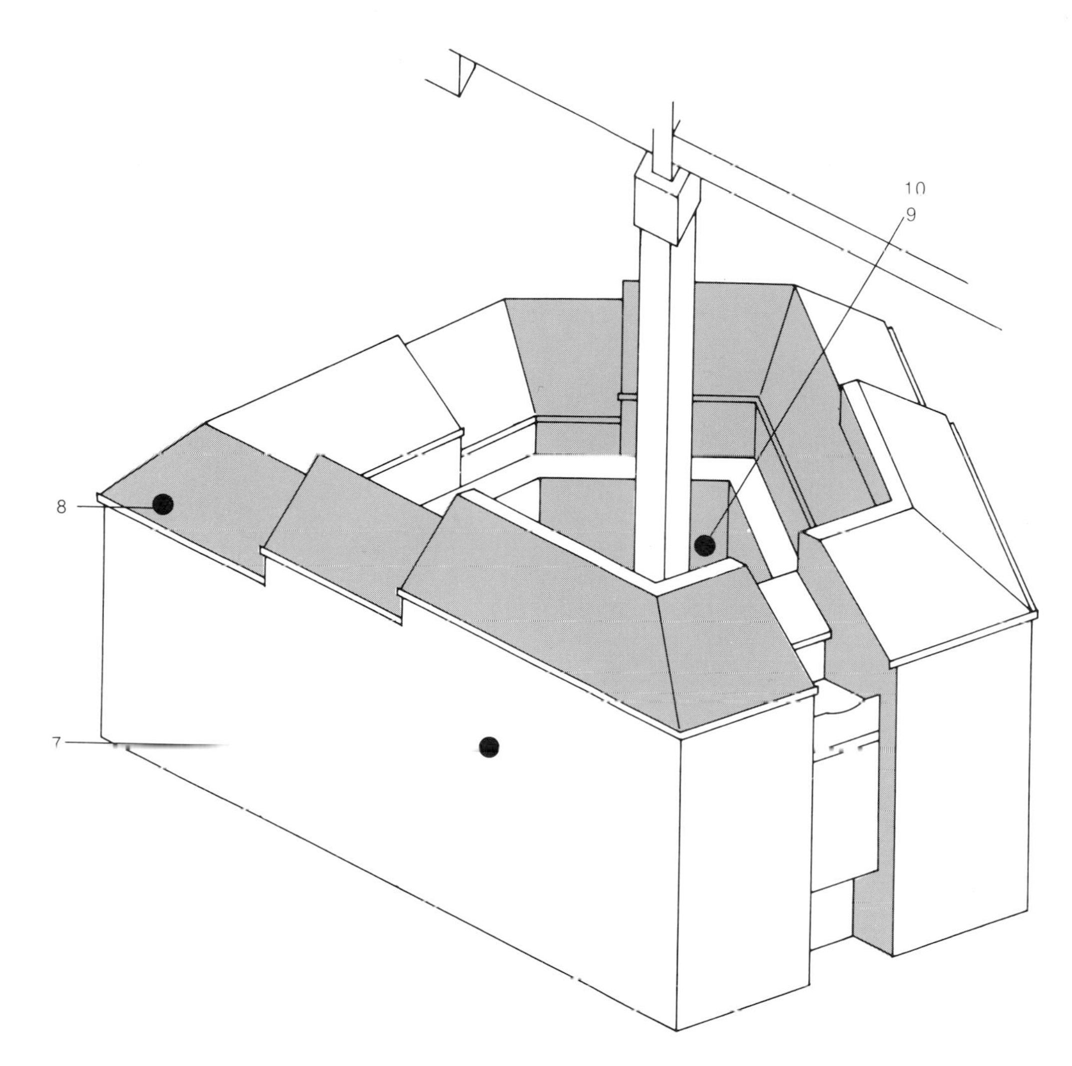

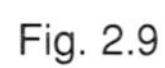

Fig. 2.9

ORGANISATION – CONSULTATION (Fig. 2.9)

Costs arising from organisation errors	Prevention costs

7. *Market study – dimension of bays*

The dimensions of window bays were determined without knowing the outcome of the tenders for 'external carpentry'. Once the main building works had begun, it proved impossible to utilise a standard frame which was particularly well designed, and which was offered at a good price.

Consequences:
Extras cost arising from fabrication to measure
5,980 ECUs

Setting parameters beforehand and securing better co-ordination at the tender stage would not have cost any more

8. *Lack of consultation – main works – structural framework*

A detail in the architectural plans called for a special link between the masonry and the structural framework to secure a particular type of insulation. A variant proposed by the roofing contractor, through a modification in one detail of the main building works, turned out to be technically more advantageous as well as being less costly. The failure to communicate this modification to those in charge of carrying out the main works brought increased labour and loss of time in order to resolve a conflict.

Consequences:
432 ECUs

Properly transmitted information would not have cost extra

9. *Imprecise specifications*

The preference shown by the commissioning client for joint contracting between companies meant that there was a need to define in contractual terms the arrangements for utilising energy, hoisting equipment, site installations, lighting in shared areas, removal of waste, and worker protection.
As this point was not made clear in the specifications, conflicts arose between companies and were negotiated during the site work:

- site malfunctions involving an unquantifiable reaction by the commissioning client;
- three consultation meetings involving seven people;
- designation of a procedure, paid for as an administrative cost by the commissioning client (2,760 ECUs)

Dealt with at the right time, the contractual organisation of the companies would have cost a half-day of administrative work and reflection
184 ECUs

Costs arising from organisation errors

10. *Special techniques*

Without the communication of this information, the costs of on-site assembly of the installation, delivered and dismantled on the spot, would have risen to 391 ECUs

Fig. 2.10

Prevention costs

The heating company made it clear, when its tender was accepted, that the opening designed in the roof to receive the type of installation ordered was too narrow (Fig. 2.10). As the main works had still only reached the fourth storey, it was possible to modify the plans without any extra implementation cost.

Drawing up the detail:
34 ECUs

OVERALL CALCULATION OF THE COSTS OF POOR QUALITY AND THE COSTS OF PREVENTION IN CONSTRUCTING A BUILDING (Fig. 2.11)

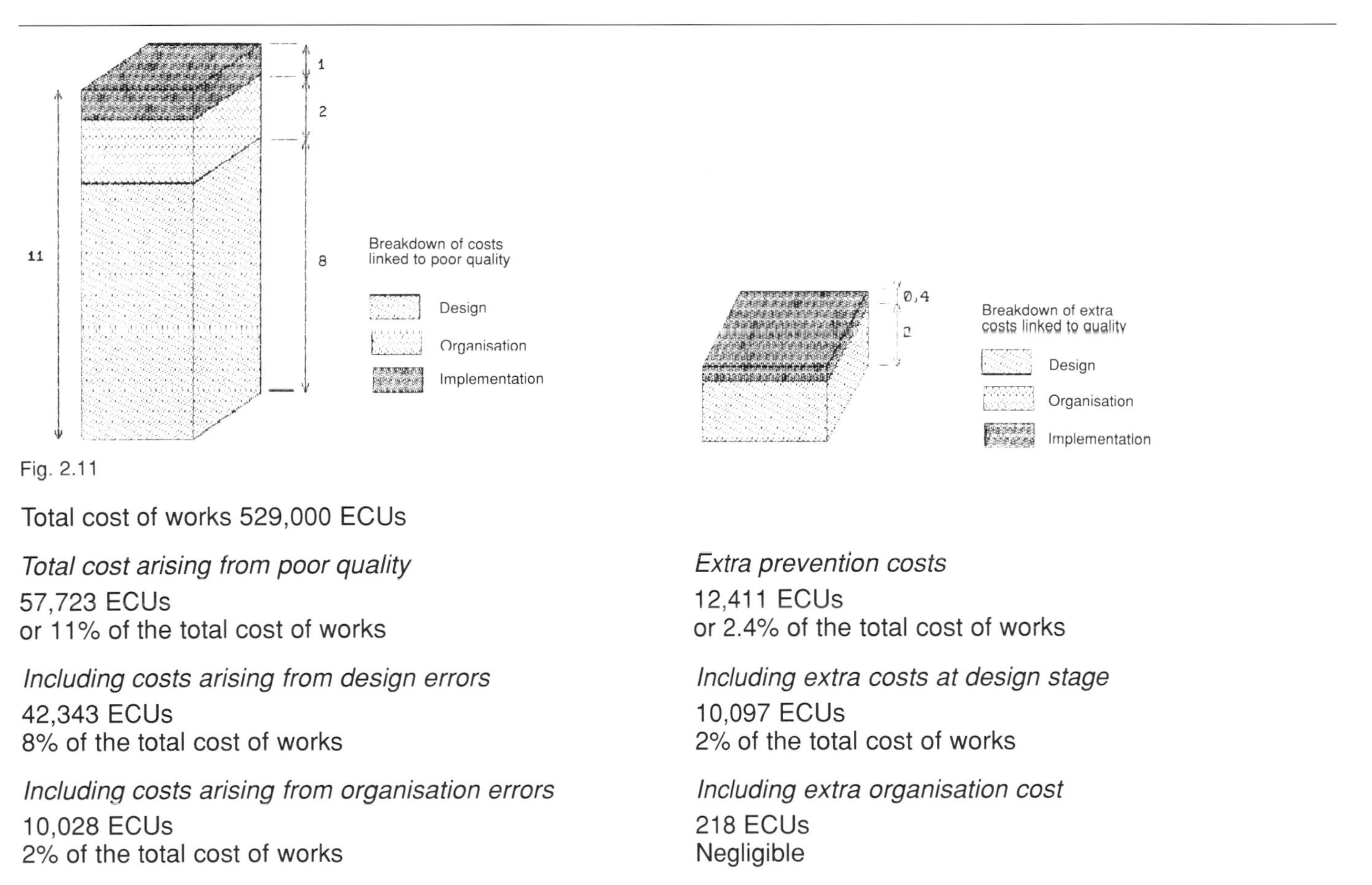

Fig. 2.11

Total cost of works 529,000 ECUs

Total cost arising from poor quality

57,723 ECUs
or 11% of the total cost of works

Including costs arising from design errors

42,343 ECUs
8% of the total cost of works

Including costs arising from organisation errors

10,028 ECUs
2% of the total cost of works

Including costs linked to errors at the implementation stage

5,817 ECUs
1% of the total cost of works

Extra prevention costs

12,411 ECUs
or 2.4% of the total cost of works

Including extra costs at design stage

10,097 ECUs
2% of the total cost of works

Including extra organisation cost

218 ECUs
Negligible

Including extra prevention costs at the implementation stage

2,096 ECUs
0.4% of the total cost of works

IMPLEMENTATION

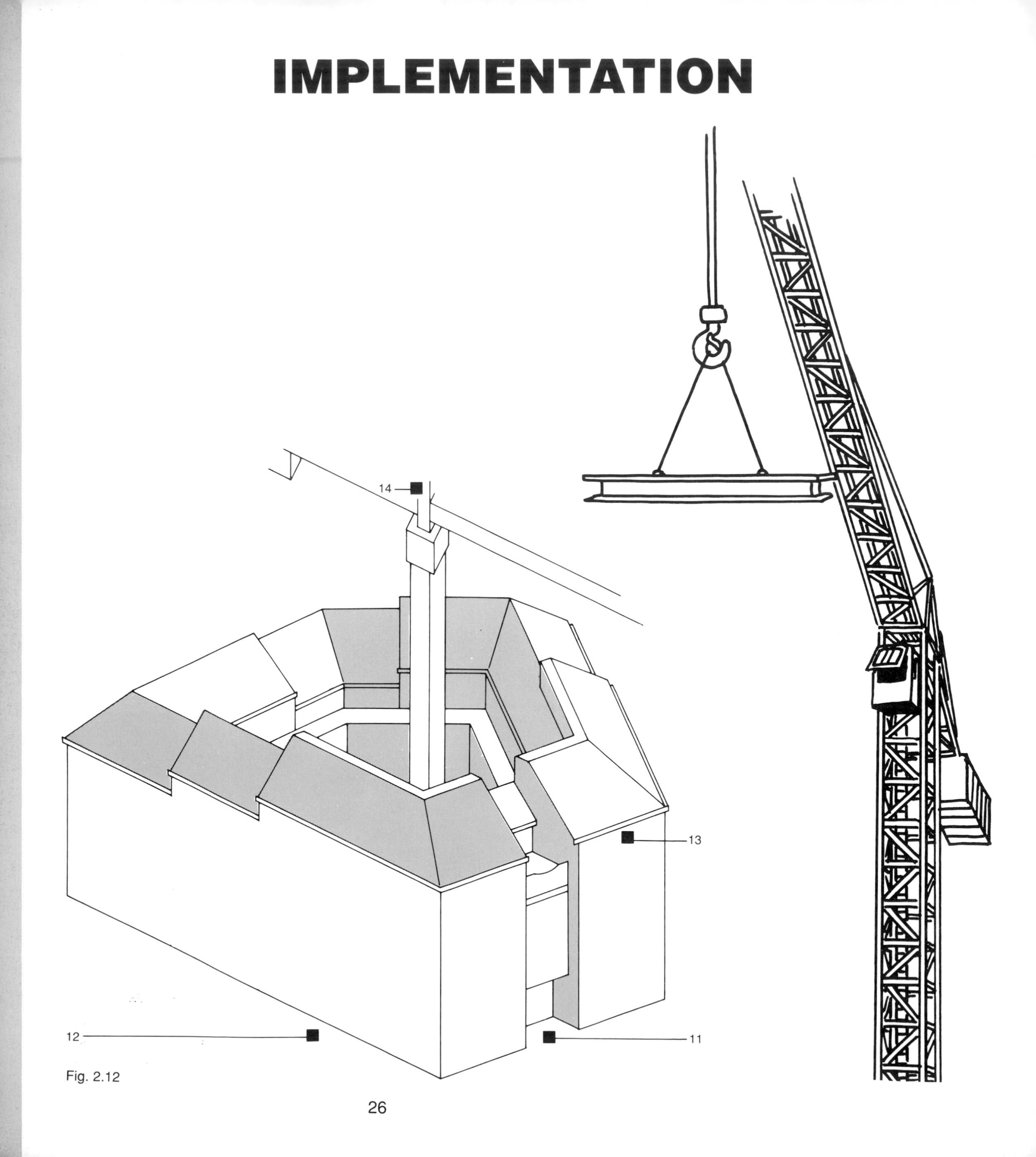

Fig. 2.12

IMPLEMENTATION (Fig. 2.12)

Costs arising from implementation errors	Prevention costs
11. *Reading error*	
Following a reading error in surveying, the depth of foundations for one concrete curtain wall went 30 cm beyond its intended dimension. Consequences: – useless work; – arguments on site; – extra concreting along the entire wall; – shuttering 527 ECUs	A first surveyor's reading checked by a second or even a third reading 3 ECUs
12. *Removal of waste*	
At the end of the main works, the waste was thrown down from the different floors on to the ground floor, and then loaded on to a lorry 3,565 ECUs	An evacuation funnel (Fig. 2.13) accessible on each floor, and leading to a covered skip on the pavement would have cost 1,955 ECUs

Fig. 2.13

13. *Installation of anchorage points*

For the safety of workers involved in the construction and later the maintenance of the building (painting the ledges, roofing, placing roof frames), the architect designed anchorage sockets for the positioning of brackets, at the request of the commissioning client (Fig. 2.14)

Consequences:

As they were not positioned when the girders and pillars were being concreted, it was necessary to drill through the concrete in order to allow them to be fixed.
Cost: 690 ECUs

Nothing

Fig. 2.14

14. *Protection of the crane mast*

The mast of the crane leased by the company was stained by concrete spillages when floors were being poured.

Consequences:

The owner of the crane, after dismantling it, invoiced the cleaning costs.
1,035 ECUs

Cost of a plastic sheet to protect the mast of the crane
138 ECUs

2.2 EXTENT AND COSTS OF OCCUPATIONAL ACCIDENTS

2.2.1 INDICATORS

2.2.1.1 Frequency of accidents

Double the average for other sectors in the economy (Fig. 2.15).

2.2.1.2 Degree of gravity of accidents

Triple the average for other sectors in the economy.

2.2.1.3 Fatal accidents in the sector

One-third of the total for the entire industrial sector (8).

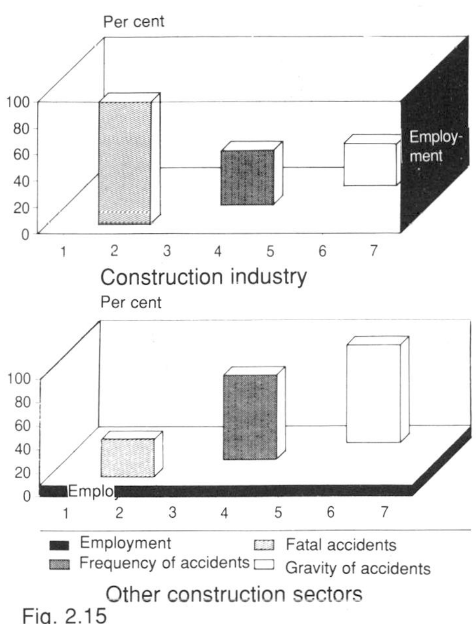

Fig. 2.15

2.2.2 COST OF ACCIDENTS

2.2.2.1 The cost of occupational accidents in construction

This represents 20 per cent of the cost of occupational accidents for the entire industrial sector, with a proportion of employment standing at about 7 per cent.

2.2.2.2 The cost of accidents – 3 per cent of the turnover in the sector

The total cost of accidents (insured or otherwise), their direct costs (salary and rehabilitation costs) and indirect costs caused by the accident (replacement of the victim, breakage of equipment, loss of production, etc.) represents 3 per cent of the turnover in construction and public works, rising to between 7 and 10 per cent of the salary bill in the sector, leaving aside the cost of the human life involved (4, 10).

2.2.2.3 The cost of worker protection: 1.5 per cent of turnover

The cost of preventing occupational accidents within companies, on the basis of a strict application of regulations relating to collective protection represents:

- 1.5 per cent of turnover for companies doing main building works;
- 0.4 per cent of turnover for companies involved in finishing;
- 5 per cent of turnover for companies erecting structural frames and roofing.

This means an overall cost for collective protection,* taking account of the turnover in each sector of activity, of **1.5 per cent of turnover** in the construction and public works sector, or **half the cost of accidents** (10)

Fig. 2.16 This access shows: on the one hand, inadequate safety measures on the part of the roofing company; on the other, failure on the part of the designer to take account of the constraints of site work and later maintenance; the designer has simply forgotten to provide access to the roof.

2.2.3 CAUSES

2.2.3.1 Decisions upstream of the site

An analysis of fatal accidents on building sites tends to show that about two-thirds of them are due to shortcomings in design (architectural choices, decisions on materials and equipment (Figs 2.16 and 2.17) and organisational problems (especially the consequences arising from combined activity by members of different trades) (4).

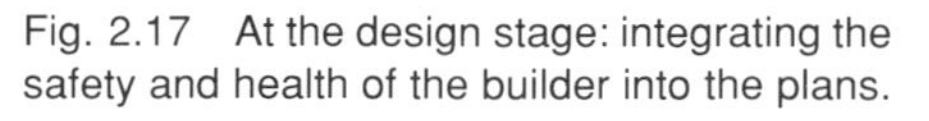

Fig. 2.17 At the design stage: integrating the safety and health of the builder into the plans.

* The concept of 'collective protection' covers the equipment used (guard-rails), the operational mode (safety in the workplace), general equipment (extinguishers, stretchers), but also the cost of various checks on material (cranes, electric boards, etc.).

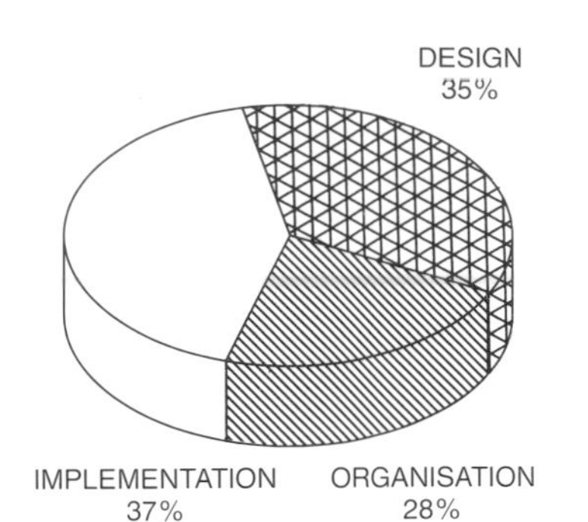

Fig. 2.18

Fig. 2.19

Design Thirty-five per cent (Fig. 2.18) of fatal work accidents in construction are caused by falls (Table 2.3 and Fig. 2.19). These can be diminished mostly through **architectural design** and the **design of equipment and material and work stations**.

Organisation Twenty-five per cent of fatal accidents arise from performance of simultaneous but incompatible activities. The planning process must therefore:

- take account of these incompatibilities (e.g., not carrying out soldering operations in the presence of painters using volatile products, Fig 2.20);
- deciding not to exceed a certain workforce load: an irregular and high work rate increases the risk of accidents.

Implementation The remaining 37 per cent of fatal accidents are attributable to risks on the site, company equipment (unsuitable, badly maintained), working conditions in the companies, the circulation of people and materials on the site, and training policies within companies.

Table 2.3 Accidents following falls from heights (11): 35% of fatal accidents; 30% of total cost of accidents in the sector

Main sources	Scaffolding, shuttering	Roofing, facades, glazing, maintenance	Finishing works
% of accidents following falls from heights	31%	18%	12%
Prevention	Equipment design and organisation	Architectural design	Organisation

We may therefore conclude that about 60% of fatal accidents on building sites arise from decisions made upstream of the site!

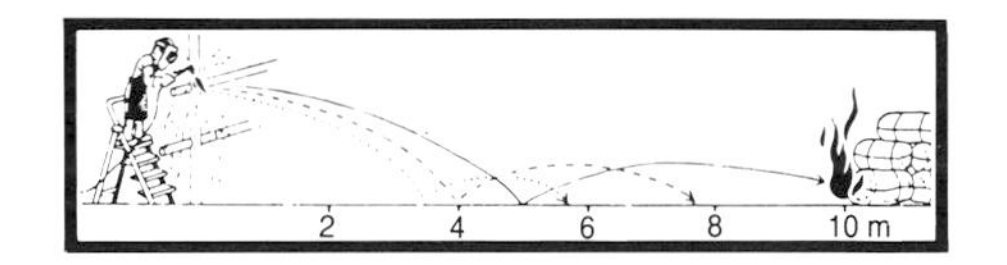

Fig. 2.20 Spillages of incandescent particles are dangerous up to a distance of more than 10 m.

2.3 COST OF INFERIOR WORKING CONDITIONS

The improvement of efficiency in an organisation must include the improvement of working conditions.

2.3.1 INDICATORS

The indicators for evaluating these factors include the following:
- failure to respect deadlines;
- useless salary costs;
- excessive consumption;
- deviations from expected productivity (2).

2.3.2 COSTS

An analysis of social and economic costs carried out in construction companies reveals that:
- absenteeism represents 22 per cent of the salary bill;
- occupational accidents represent 8.4 per cent of the salary bill;
- staff turnover on sites affects 30 per cent of personnel;
- remedial work on site represents 7 per cent of the salary bill;
- productivity shortfalls represent 17 per cent of the salary bill (2).

2.3.3 CAUSES

The analysis of the causes of malfunctions shows (Table 2.4 and Fig. 2.21):
- that there are no simple 'cause and effect' relationships, but rather a combination of factors establishing links between them;
- that there are cumulative malfunction effects, from one indicator to the next;
- that these effects are passed on from design to implementation, and from company to company.

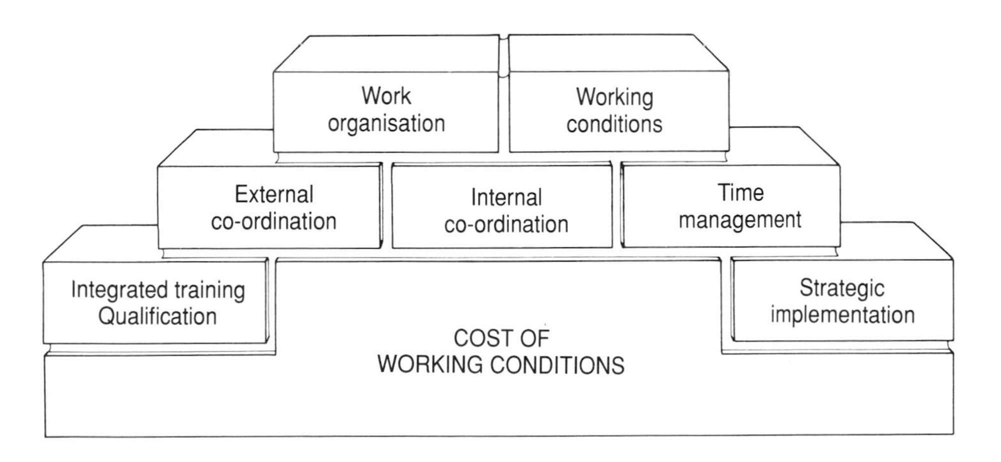

Fig. 2.21

Table 2.4: Summary of causes of malfunction under major headings. The cost of working conditions is the outcome of their combination

Type of causes cited	Description
1. Working conditions	The physical setting (cold, heat, noise, etc.). Physical load. Nervous tension. Context of relationships. Extra hours worked
2. Work organisation	De-motivating monotony of certain processes. Division of work excessively marked. Unsuitable rates of work. Defective transmission of information. Imprecise definitions of functions
3a. External co-ordination	Imprecise studies carried out too late. Preparation timescales too compressed.
3b. Internal co-ordination	Insufficient contact between hierachical managers. Inadequate preparation of work. Insufficient contact between management and shop-floor. Defective communication of the criteria to be applied. Badly defined quality criteria
4. Integrated training/ qualification	Unsuitable initial training. In-service training not followed by practical application. Lack of training for certain techniques and equipment
5. Time management	Tasks inadequately accepted by supervisors. Dispersal of middle management tasks. Time management at senior level
6. Strategic implementation	Certain tasks considered secondary in relation to objectives. Clarification of objectives. Definition of resources matching objectives. Remuneration and qualification levels

Source: Adapted from ISEOR, January 1986 (2).

CHAPTER 3
COSTS USUALLY ORIGINATING UPSTREAM OF THE SITE

After having tried to outline the scale and structure of costs arising from poor quality, from occupational accidents, and more especially from difficult or dangerous conditions for doing the work, it is necessary to identify more clearly the different functions involved in controlling these costs.

Despite appearances, the objective is not to engage in excessively detailed sequencing of the different phases occurring between the original design and the final completion of the project.

The ultimate objective is actually the opposite, because, as we shall see later (Chapter 4), excessive pigeon-holing and sequencing would militate against the achievement of a new rationale covering the whole process, an essential condition in order to secure fresh productivity gains in the sector.

However, for the purposes of demonstration it has been found easier to group these functions under three categories:

- design functions (or phases);
- organisation functions (or phases);
- implementation functions (or phases);

This three-part division also allows the possibility of bringing out one essential factor: the site phase in the strict sense (implementation) is only one of the phases (or one of the places) where improved productivity and quality can be sought. There can be an excessive tendency to seek productivity gains at this later phase only. All too often, when that perspective is adopted, the site phase is assigned the function of correcting the errors and delays created upstream, which must now be made good. This is particularly true of ocupational accidents which, as we have seen, often result from decisions made before the site phase. All the same, this does not exonerate companies from their responsibilities and duties to protect workers on site.

The first exercise therefore sets out to identify the functions involved in controlling the costs already mentioned.

Table 3.2 Tables and graphs summarising the findings of the survey carried out by the Labour Ministry in October 1978. The presentation is based on the methods proposed by Professor Roustang of LEST in Aix-en-Provence

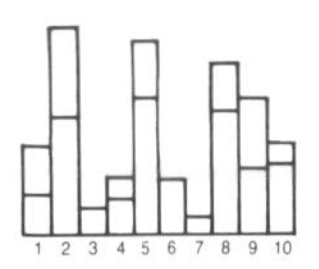

	Construction only (%)	All activities (%)
Work environment		
1. Temperatures often too high or too low	30.5	14.3
2. Frequent draughts	73.0	43.1
3. Too much or too little light in the work position	6.8	8.7
4. Breathing in smoke	11.6	20.0
5. Breathing in dust	68.5	47.7
6. Contact with toxic products	19.1	20.9
7. Contact with explosive products	5.5	6.3
8. Working in 'filthy conditions'	61.0	44.6
9. Working in damp conditions	47.8	23.7
10. Experiencing disagreeable smells	26.7	32.5

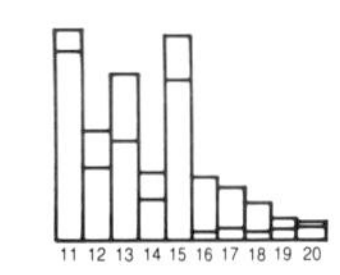

	Construction only (%)	All activities (%)
Work-load		
11. Have to stand for long periods	74.4	66.5
12. Uncomfortable or tiring work posture	38.9	26.4
13. Have to carry or move heavy weights	58.6	35.1
14. Suffering vibration or shaking	22.6	14.2
15. Working more than 40 hours per week (1976)	71.2	54.3
16. Doing shiftwork	1.4	20.4
17. Working under extreme time contraints	3.3	17.5
18. Repetitive work	1.4	12.9
19. Cannot speak to each other on account of the work	2.4	6.8
20. Are isolated in their work positions	4.0	6.4

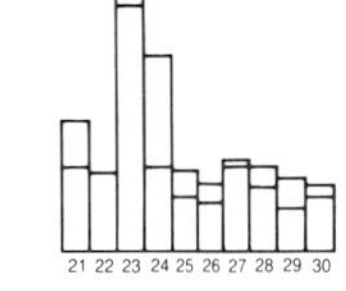

	Construction only (%)	All activities (%)
Perception of risks and nuisances		
21. Cannot speak to colleagues	29.1	44.5
22. Cannot interrupt their work	27.2	29.9
23. Feel exposed to all sorts of accidents	95.0	86.4
24. Risk of serious fall	68.2	29.7
25. Risk of electrocution	27.4	19.1
26. Risk of being burned	16.3	23.5
27. Risk of being injured on machinery	29.2	33.6
28. Risk of traffic accidents	29.2	24.7
29. Risk of other accidents	24.6	14.9
30. Risk of breathing toxic substances	18.6	22.5

3.1 IMPORTANCE OF THE THREE PHASES OF THE PROJECT AS REGARDS THE QUALITY OF LIFE AND WORK ON THE SITE

- First function – design;
- Second function – organisation;
- Third function – implementation.

3.1.1 A PERCEPTION OF 'POOR QUALITY' DIFFERING FROM UPSTREAM TO DOWNSTREAM IN THE DECISION-MAKING PROCESS

Surveys assembled in the framework of the present study reveal that (Table 3.1):

- 80 per cent of damage to works and malfunctions on site may be imputed to management errors which are distributed across the upstream and the downstream phases (design, organisation, choice of equipment, etc.);
- 20 per cent are due to errors during the implementation phase proper (12).

Table 3.1

Source of damage	Belgium 1974–5 (%)	Belgium 1976–7 (%)	England 1970–5 (%)	Germany 1970–7 (%)	Denmark 1972–7 (%)
Design	49	46	49	37	36
Implementation	22	22	29	30	22
Defective materials	15	15	11	14	25
Mistakes in use	9	8	10	11	9
Miscellaneous	5	9	1	8	8

The perception of these costs and inadequacies varies from individual to individual, according to the hierarchical position occupied by different people within the entire process (13).

The results of two surveys show that workers have a definite awareness of the positive impact on their working conditions which may be produced by decisions taken before the site phase. They often view those conditions as unpleasant and dangerous. But at the same time they value the autonomy which they have in doing their work (see Table 3.2).

The other survey shows that designers are very unaware of the impact which may result from their decisions taken before the site phase regarding working conditions and safety for the workers.

3.1.2 IMPORTANCE OF CO-ORDINATION OF OBJECTIVES AS THE PROJECT ADVANCES

It is therefore important to define the following aspects from the design phase of the project.

3.1.2.1 The objectives associated with the constraints set by the partners from the project orientation phase

3.1.2.2 The functions belonging to the project, the users and the builders

Control over economic and social costs is shared between the upstream and downstream phases when objectives are being co-ordinated between:

– the commissioning clients and the designers;
– the designers and the surveying officers;
– the main contractors and the employers;
– internal and external services within companies;
– subcontractors;
– main building works and various portions of finishing works.

This shows the importance of liaison between the different participants and the different sequences.

3.1.2.3 Exposure to bearing the consequences of inadequacies attributable to other partners

This risk must lead everyone involved to try taking preventive action through a clear statement of each party's objectives and constraints, and through an effective programme of information on constraints and objectives arising further downstream (Table 3.3 and Fig. 3.1).

Table 3.3 *Examples*: defective communication between the participants

A worker

A worker who discovers an error of design or organisation while carrying out repetitive tasks (boring, height of thresholds, installing heating equipment, etc.), will not inform those responsible upstream – because they seem inaccessible – so that arrangements could be made to correct those errors during the ongoing works preceding his own intervention

Site supervisor

A site supervisor finding a gain during concreting works and a loss during masonry works, arising from nonconformity of materials and the demand by the labour inspector that the safety of all work posts should be reviewed (scaffolding, guard-rails, flooring, etc.) will try to provide the site management with a balanced budget report

Site manager

The site manager will use these 'positive' conclusions in drawing up subsequent reports which will be presented to the clients

Designer

This customer, the designer, will use these unreliable data in later budgeting other projects, surveyed for hypothetical commissioning clients, before consulting companies

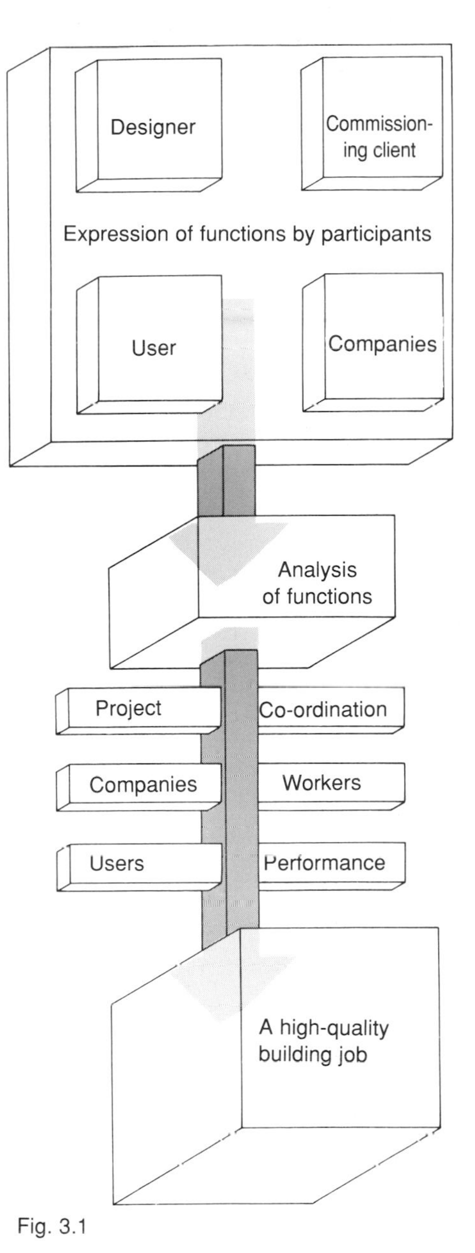

Fig. 3.1

3.2 EXAMPLES OF ENHANCEMENT

3.2.1 AT THE LEVEL OF DESIGN

- architectural design;
- equipment design;
- materials design;
- work post design.

Controlling site hazards, identifiable during the project design phase, is the way to build safety into the project.

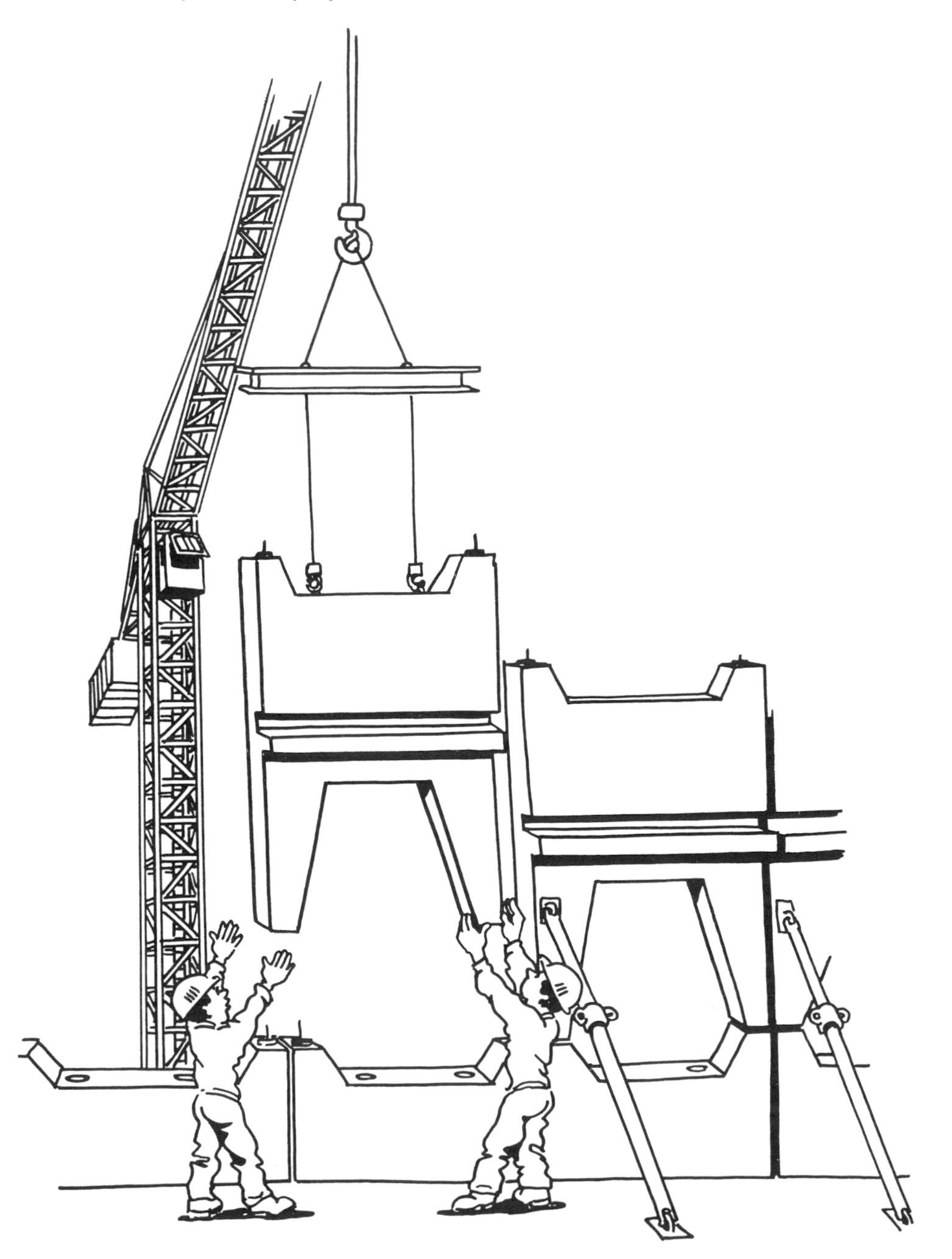

3.2.1.1 Architectural design

The prevention of falls from heights should be part of the construction procedures. Architectural design can bring considerable improvement to working conditions on sites.

Fig. 3.2

All too often, when a building is being constructed, protection against falls is only introduced when the dangerous operations have already been completed. As an example, take this guard-rail (Fig. 3.2) which makes it appearance after the shuttering, metalwork and cementing operations, where the risks are greatest.

In producing a type of facade which serves the same purpose, the risks of falls may either be innumerable or else be almost completely eliminated. The architectural detail shown by Fig. 3.3 involves a maximum risk of falls. To protect workers under these conditions requires the use of sophisticated procedures which do not form part of the assembly process itself.

A differently designed detail for this facade, producing the same end result as the previous example, uses an architecturally designed concrete feature which forms the support for the floor and the upper portion of which serves as a permanent guard-rail. The protection against falls makes its appearance before the worker has been exposed to this hazard (Fig. 3.4).

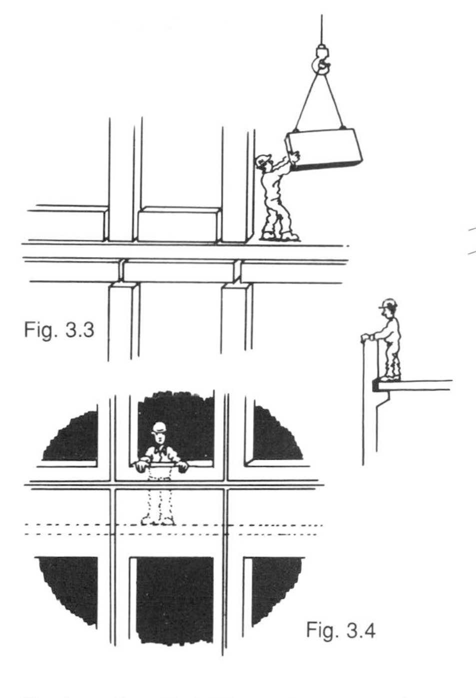

Fig. 3.3

Fig. 3.4

During the finishing phase, various hazards are present (Fig. 3.5):

- a guard-rail not replaced;
- holes bored for the insertion of ducts in the beam supporting the floor above (often the site works are already in progress while the special technologies have not yet been defined);
- filling in the window bay.

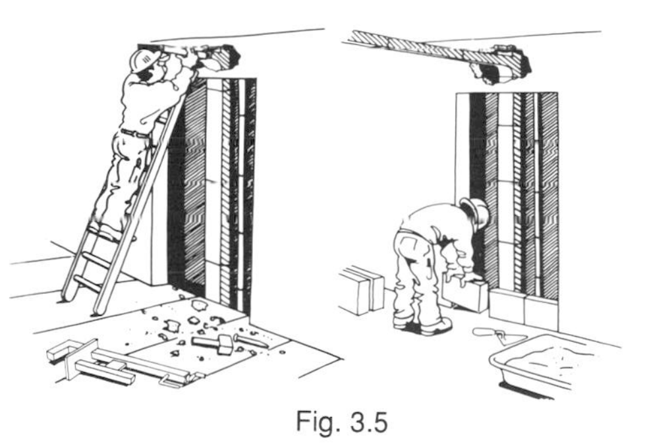

Fig. 3.5

On the other hand, the following situation is caused by architectural decisions made before the site phase. This inverted beam protects the worker and leaves a passage free for ducts in the ceiling. The window bay can be filled in without hazard (Fig. 3.6).

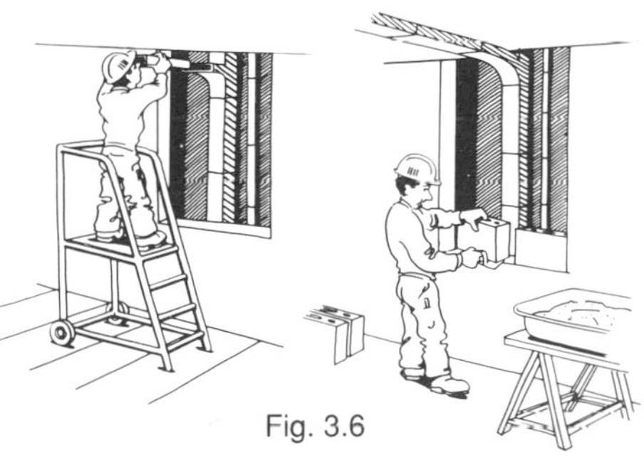

Fig. 3.6

3.2.1.2 Equipment design

These bracket anchorage points will be useful both for the construction of the building and for its maintenance. They will also be useful when it is being demolished (Fig. 3.7).

Scaffolding made of tubes and connections, leased with optional handrails and working floor, has now almost disappeared from the market, to be replaced by scaffolding where stability is assured through the flooring and guard-rails (Fig. 3.8).

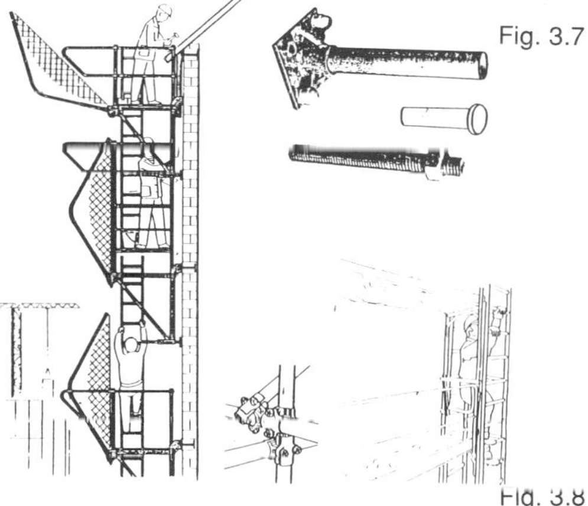

Fig. 3.7

Fig. 3.8

The excessively frequent use of pneumatic drills, for example, which is a nuisance both for workers and for local residents, can arguably be traced back to errors in the original design (Fig. 3.9).

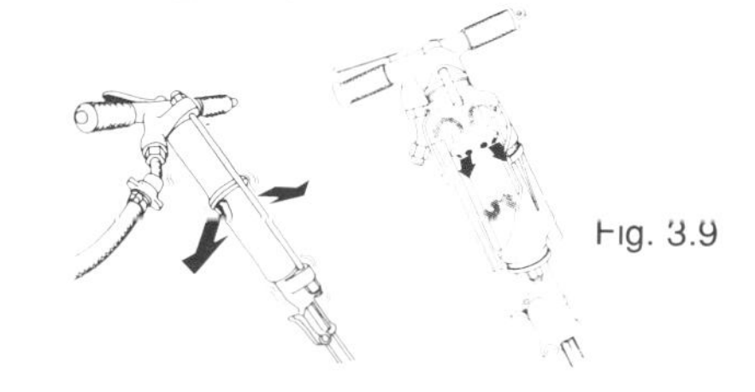

Fig. 3.9

3.2.1.3 Materials design

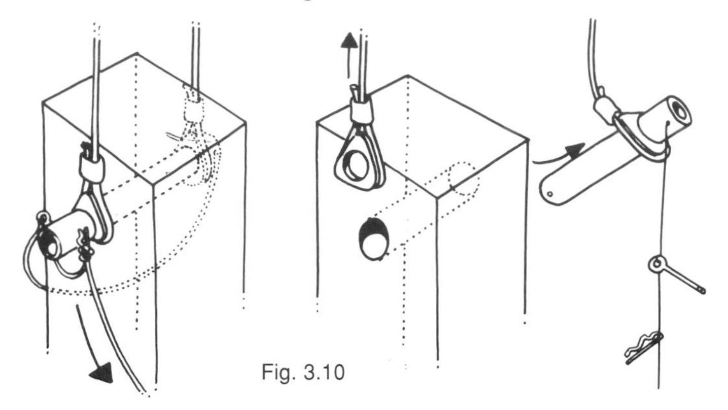

Fig. 3.10

When prefabricated centres are being erected, accessories for remote-control unhooking (Fig. 3.10) provide an improvement in working conditions:

- no more ladders resting on an uneven floor – as many of these as there are pillars on the site;
- performance, comfort and safety are all linked together here.

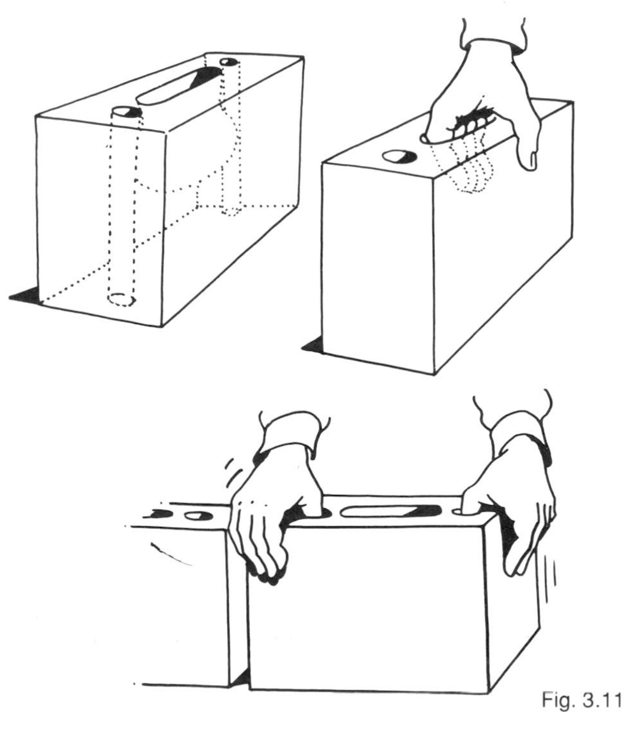

Fig. 3.11

An ergonomic study of bricklaying work has shown that the work-load of a bricklayer could be considerably reduced (20%) by inserting a space in the building block to allow the hand or the thumbs to be inserted to grip the block (Fig. 3.11). Work performance increases by some 17 per cent. This also greatly reduced spinal strain and increased worker satisfaction and the quality of work.

3.2.1.4 The design of maintenance equipment

Maintenance – servicing Some oil-burners require the worker to have another pair of hands, in order to link the transmission axis (which is not all of a piece with the motor) to the ventilator (two hands), with the cap of the ventilator being simultaneously bolted to the bed-plate (another hand). Other types of equipment, fortunately, allow immediate linkage and only require the use of one hand (Fig. 3.12).

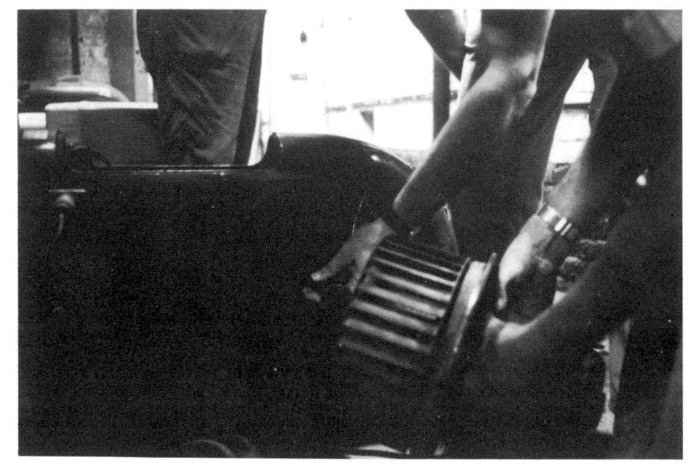

Fig. 3.12

3.2.2 ORGANISING THE WORKS

Upstream from the site and within the companies, good organisation of the works exists only where it has been envisaged right from the earliest stages of project orientation (Fig. 3.13).

Fig. 3.13

The second of the three stages, 'Organisation', falls into two parts.

3.2.2.1 *The architect*

The architect is the professional person who advises the commissioning client, setting the lead times by planning the order in which different building professions will come on site.

Avoiding incompatible tasks The architect's planning will indicate dangerous phases of work, making sure, for example, that the heating installer will not use a blowpipe close to a painter or volatile substances (Fig. 3.14).

Shared equipment on the site Figure 3.15 shows equipment used on a shared basis by companies. Given an identical timescale, working conditions will be totally different depending on whether or not organisational ergonomics and safety considerations are built into the planning process.

Maintaining a constant level of employment as far as possible Good planning consists in flattening out the curve of employment, allowing the site staff to be increased at certain stages in order to make good any possible delays (Fig. 3.16).

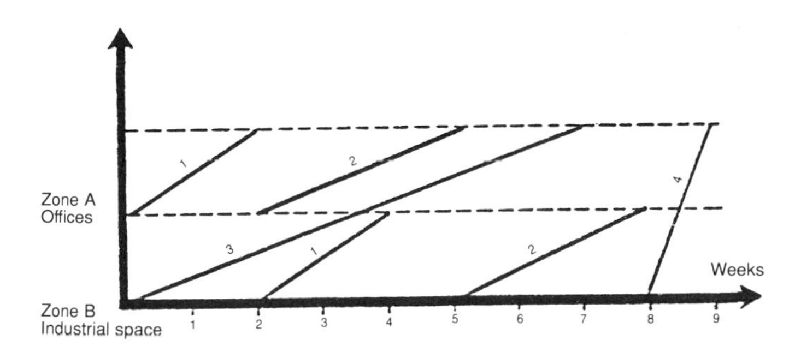

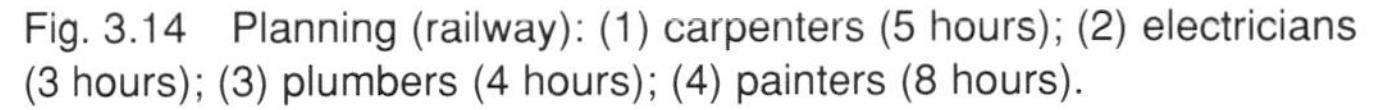

Fig. 3.14 Planning (railway): (1) carpenters (5 hours); (2) electricians (3 hours); (3) plumbers (4 hours); (4) painters (8 hours).

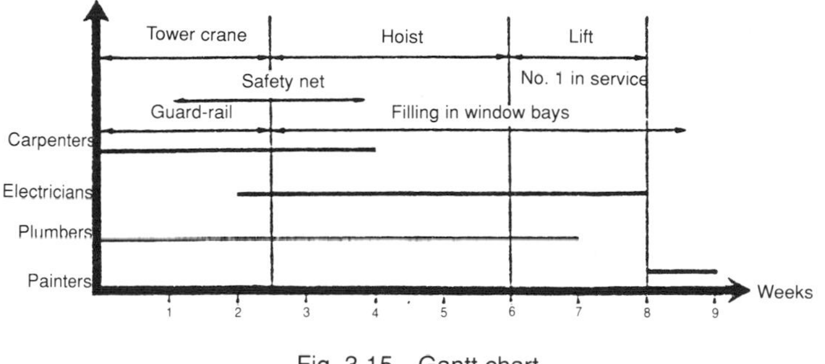

Fig. 3.15 Gantt chart

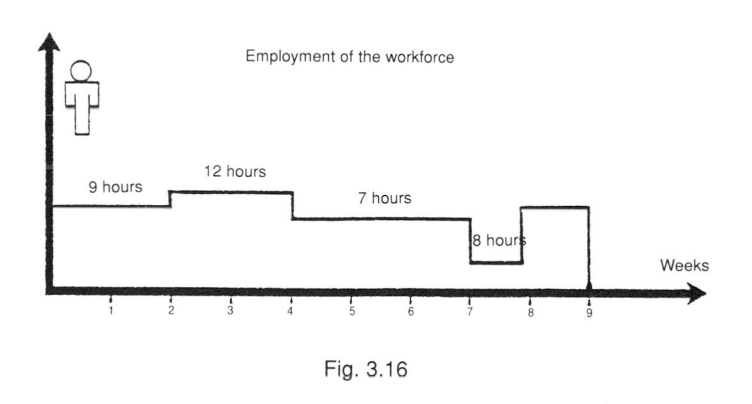

Fig. 3.16

Well-designed planning, both on the part of the architect and on the part of the employer, improves working conditions on the building site.

Movements on site The majority of accidents involving falls (from heights or on the level: 30 per cent of total cost of accidents) happen when workers are moving around the site (unprotected openings, ladders, badly assembled scaffolding, etc.) (Fig. 3.17). Depending on the site and its organisation, the time spent in moving from place to place can be increased up to 100 per cent. The organisation of site movements therefore takes on crucial importance in the area of productivity and safety.

Fig. 3.17

Roofing over Planning the roofing over of the works in the shortest possible timescale helps to avoid the risk of materials deteriorating due to bad weather; it also improves working conditions.

Fig. 3.18

Access routes Avoid intersections between vehicles on site, or vehicles undertaking dangerous reversing procedures (Fig. 3.19).

Fig. 3.19

3.2.2.2 The employer

The planning undertaken by the employer in charge of the main works will specify, on a chart, the areas reserved for:

- site installations;
- circulation;
- storage of materials, etc.

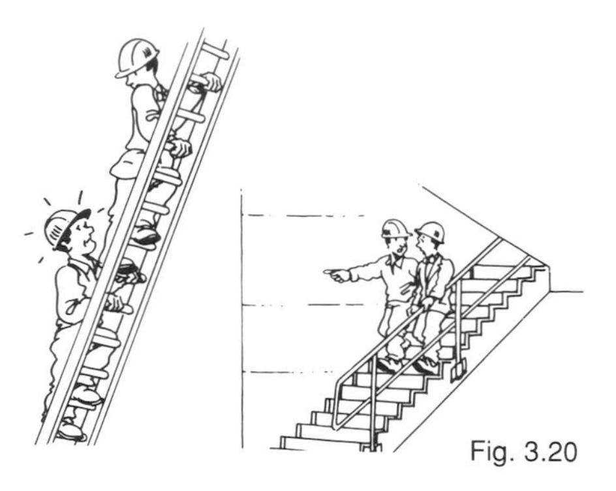

Fig. 3.20

Immediate installation of stairways Good planning also brings considerable improvements to workers' comfort. Thus, the work-load involved in setting up a ladder between one storey and the next is three times higher than the effort needed to go up a stairway. The time spent on the ladder is also much longer (Fig. 3.20). By planning the immediate installation and railing of stairways, one avoids dangerous handling in stairwells.

3.2.3 CARRYING OUT THE PROJECT

3.2.3.1 Choice of suitable equipment and collection of information from the workers

The third stage in our three-part 'site' analysis is located within the individual building company. By taking action on the choice of equipment and by listening to the workers, ergonomic action can considerably improve the quality of working conditions.

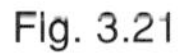

Fig. 3.21

If one places these protruding screws 20 cm higher, they are no longer dangerous to the worker's eyes (Fig. 3.21).

Unlike a carelessly nailed batten, this removable handrail fits across lift shaft entrances without damaging the finishing works (Fig. 3.22).

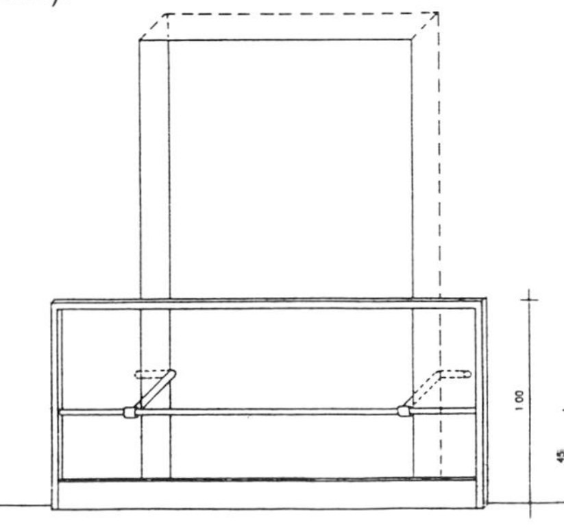

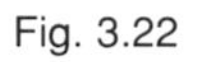

Fig. 3.22

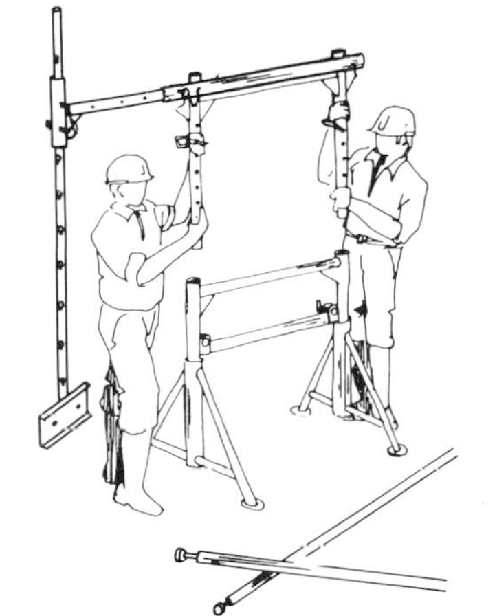

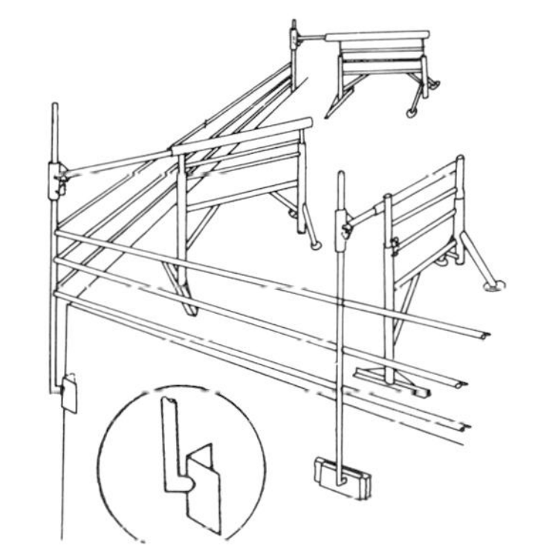

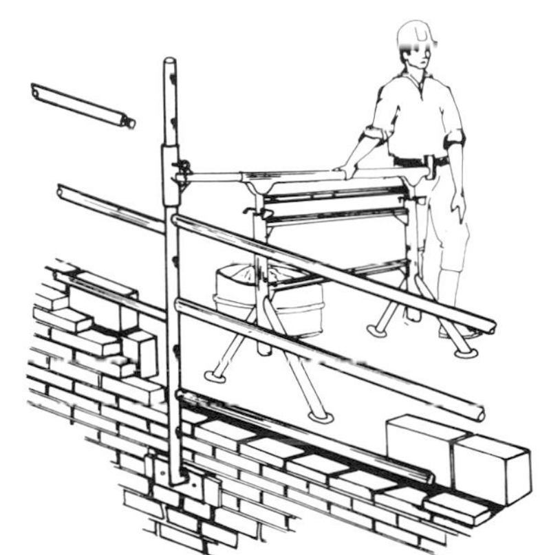

Fig. 3.23

Protection of bricklayers This is achieved by an extendable handrail (Fig. 3.23). During the works:

- moving the handrails;
- raising the mobile part of the trestle.

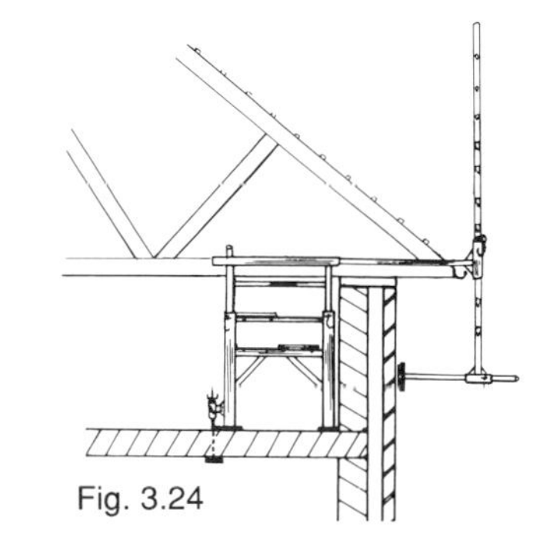

Fig. 3.24

Protection of roofers (Fig. 3.24)

- installation of trestle before the roof beams;
- stabilisation by anchorage clamp;
- installation of protective equipment.

Fig. 3.25

A high-level team comprising a laboratory director, a professor of soil mechanics, and several civil engineers was charged with preparing an advance plan for a solid pier for a marina. In the end, it was the observations and proposals made by bulldozer drivers which played the greatest part in producing the decisions determining the organisation of the main construction works (Fig. 3.25).

The best choice of ergonomic equipment can only be achieved by taking account of users' opinions.

CHAPTER 4
NEW APPROACHES TO PROJECT ENHANCEMENT

The three approaches of quality, ergonomics and value analysis are applied on a priority basis, and with increasing success, in high-performance industrial sectors.

They are based jointly on **communications** between the participants (the users, the creators and the implementers) using **crossed information channels** which have all proved applicable in the area of improving working conditions, productivity and product quality.

The object of the present chapter is to make these approaches better known in the sector, so that everybody, whether upstream or downstream, can enhance the approaches being adopted.

4.1 A NEW APPROACH TO QUALITY REQUIRING IMPROVED RELATIONS BETWEEN THE PARTICIPANTS

Everybody, at whatever level, must have the same definition of quality, the aim being to attain as widespread as possible an understanding of the concept, both upstream and downstream of the project (14).

It should be noted, however, that any approach to quality can only be validly implemented on the basis of personal commitment by the project managers, and a mobilisation of all participants in the building process.

Quality means that the product conforms to the needs expressed

- by the commissioning client;
- by the designers;
- by those who carry out the design;
- by the users of the site, whether they are workers, residents or maintenance workers;

and involving a commitment by suppliers

- users;
- implementers;
- designers;
- commissioning client.

The definition proposed demands that:

- the commissioning client,
- the designer,
- the implementer,
- the user

should each, in turn, carry out the role of supplier or client at the following stages:

- the agreement of the plan;
- the translation of the client's expressed needs into a list of specifications;
- the issuing of a tender which conforms to these criteria.

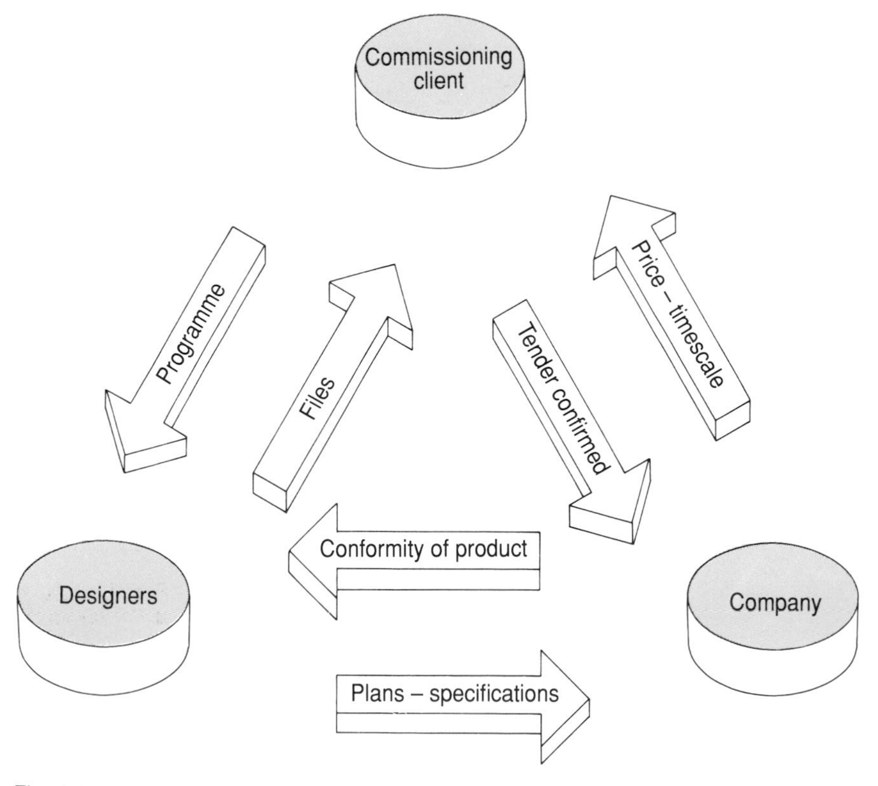

Fig. 4.1

The commissioning client expresses a need formulated in terms of a plan which the designer (supplier) translates into **plans and specifications**. These will define the product which the company (supplier) is charged with carrying out in conformity with the said specifications (Fig. 4.1).

4.1.1 COMMISSIONING CLIENT – DESIGNERS

Commissioning client = Supplier

The commissioning client must be aware that he is also fulfilling the role of a supplier. In fact, it is the quality with which he expresses his needs that will ensure a good planning of the design, execution, maintenance and exploitation of the building.

Designer = Supplier

The conformity of the product to the needs expressed by the commissioning client will be all the better to the extent that the designer has managed to define the needs of the users in an exhaustive way.

All failures by the client/supplier no. 1 to express needs precisely will be translated into a mismatch between the product as anticipated and the finished product. The greater the extent of this mismatch, the greater the gap between these two expressions.

4.1.2 DESIGNERS–COMPANIES

This relationship must make it possible to reduce the areas of nonconformity by identifying, quantifying and negotiating the needs of each participant.

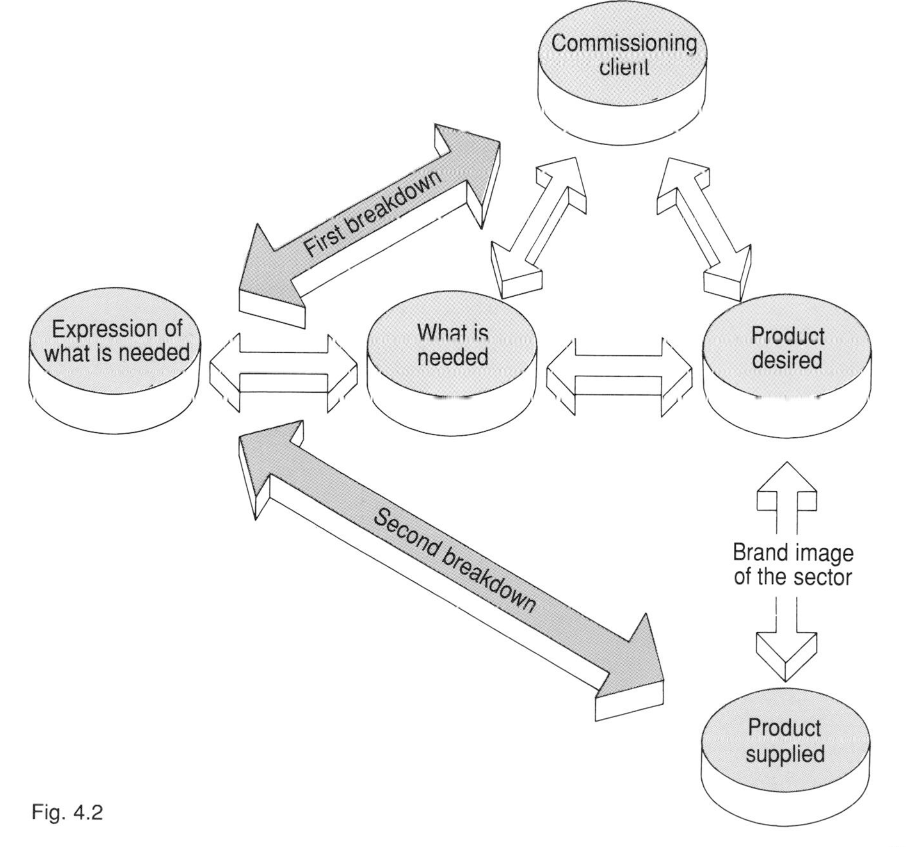

Fig. 4.2

The mismatch between the product as realised and the product desired will be all the greater as:

- the needs expressed by the participants are imprecise and badly formulated (**first breakdown**);
- the services and products supplied by these same personnel fail to conform to badly defined needs (**second breakdown**) (Fig. 4.2).

4.1.3 COMPANIES–COMMISSIONING CLIENT

The brand image of the company is affected by the quality of working conditions on its sites, among other things. **This quality expresses the quality of the finished product**.

It is therefore necessary that companies should be certain that their requirements and those of the commissioning client and the main contractor are genuinely embodied in the organisation of quality among their subcontractors and suppliers.

The organisation of quality at the implementation level alone is therefore not enough to ensure a high-quality job: it is also necessary that the planning and design should have been organised in the same spirit.

4.1.4 RELATIONS BETWEEN AND WITHIN COMPANIES (14)

A company has to deal with two spheres, as shown in Table 4.1.

Table 4.1

Supplier's sphere	Client's sphere	
Suppliers	*Internal clients*	*External clients*
Designers Subcontractors	Administrative services and sites	Commissioning client Designers
Services rendered by suppliers	*Services provided for clients*	
Plans – specifications Product design Procedures	Internal clients: equipment, techniques External clients: the building works	

Carrying out an improvement in the quality of life and the quality of the product in a company implies:

- commitment by management;
- choice of consultative management based on strategy of dialogue;
- mobilisation of all members of staff, through a permanent system of raising awareness, training, evaluation and recognition of results obtained;
- systematic measurement of:
  - client needs and the conformity of products and services to those needs;
  - the costs of quality and poor quality;
  - the satisfaction of needs expressed by workers who function alternately as clients and suppliers within the company.

4.2 ERGONOMICS

4.2.1 ERGONOMICS – A SPECIAL APPROACH

This discipline brings together all specific areas of knowledge relating to human workers and those required to design the tools, machinery and equipment which can be used with the maximum level of comfort, safety and efficiency (Fig. 4.3).

4.2.1.1 Multidisciplinary

The main feature of ergonomics is its multidisciplinary nature. In fact, the worker is at the crossroads of a number of **constraints** (noise, heat, workloads, unnatural postures, hindrances, weather, etc.) and **drawbacks** (uninteresting work, repetitiveness, delays, etc.), which have to do with very different disciplines such as occupational medicine, safety and health, applied sciences, sociology and psychology.

In addition, the ergonomic approach must be participatory and global (15).

4.2.1.2 Participatory

In effect, actual work does not necessarily match the image which decision-makers may have of it. It is therefore important to associate workers at all levels with the choice of methods and the evaluation of results.

4.2.1.3 Global

An ergonomic approach must be based on the study of all the parameters characterising the work environment, as the modification of one portion of these may introduce other nuisances or hazards. For example, anti-noise helmets may tend to isolate the workers and consequently create other hazards.

An ergonomic approach may be situated on different levels:

- design ergonomics, consisting in the advance arrangement of all parameters before a post is set up, in relation to the person or persons who will have to handle it: this is planning ergonomics;
- correction ergonomics, which consists of improving a work situation already in existence;
- organisation ergonomics, or communication ergonomics, a much broader concept covering the whole set of interrelationships within the man–machine and man–environment systems, whether one is dealing with a workshop, a building site, a service or a company.

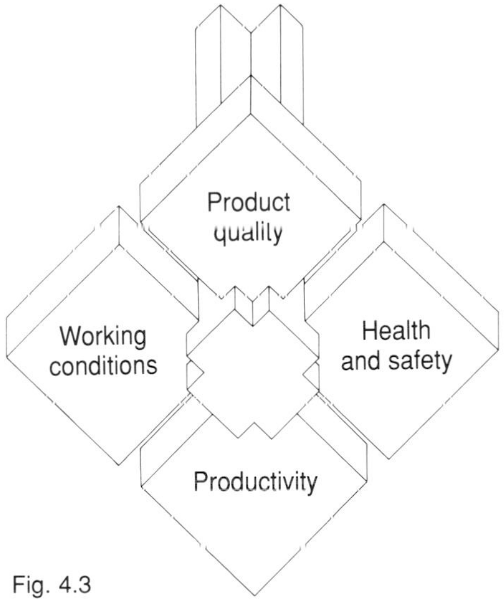

Product quality

Working conditions

Health and safety

Productivity

Fig. 4.3

4.2.2 ERGONOMICS AND INTEGRATED SAFETY ON THE SITE

Integrated collective safety is characterised by planning in advance of the building activity. Additional safety measures are those implemented when the hazards have emerged, or – worse still – after the inevitable has already taken place (15).

4.2.2.1 Examples

In the construction of a building, certain works expose the workers to risks of falling from heights. To forestall this risk, one can either surround the work surface with guard-rails, assembled on site as required, or else equip the workers with safety harnesses. These are examples of **added safety measures**.

One could also reduce the risk of falls from heights by designing work positions, access and circulation routes as far removed as possible from danger spots (holes, gaps, etc.). This is an example of safety at the level of general work **organisation**.

It is also possible to produce the necessary guard-rails in advance and plan the points where they are to be fixed, so that their installation will be easy and present the lowest possible level of risk. Better still: design a construction procedure such that the risk of falling from heights will be practically abolished. This is an example of **integrated safety**.

4.2.3 ROLES OF THE PARTICIPANTS IN DEVELOPING ERGONOMICS ON BUILDING SITES

4.2.3.1 Designer

From the earlier phases of the project, an analysis of working conditions on the site can make it possible to improve productivity and the quality of implementation of the works. In addition, the ergonomic analysis of conditions for maintaining the completed building, carried out beforehand, makes it possible to improve maintenance performance as well as the energy performance of the project. There is often a correspondence between the requirements of work on the site (productivity and safety) and the subsequent requirements regarding maintenance and service work.

4.2.3.2 The methods offices

On the basis of the documents which establish provisional costings, and refining the detail of the organisation of the works, moving from the general to the particular (general planning and planning for joint contractors engaged in the finishing work, for example), the methods office can exercise precise control over the site.

The parallel analysis of hazards connected with the different companies, their localisation in space and time, the number of workers on the site – all these allow the methods office to discuss the critical points of quality, productivity and safety with the different companies.

4.2.3.3 The workers and their company

The ability to communicate within companies, and the quality of consultation, contribute to the improvement of their competitiveness and social climate.

4.3 THE ANALYSIS OF VALUE

This makes it possible to design, redesign or research a 'product' in the broad sense, in a way which provides at the best possible cost (effectiveness) satisfaction for the user (quality) (Fig. 4.4). This approach involves the different areas of expertise simultaneously rather than successively, and brings together:

- the commercial functions;
- the research and development functions;
- the technical functions;
- the production functions

in terms of performance, to facilitate innovation at all stages of the project (16).

This is an approach which is expressed through a definition of the project in terms of objectives rather than resources.

The analysis of value consists particularly in meeting the demand at the best cost. Taking account of the current situation, this approach will make it possible to reduce the price of work done by something between 15 and 20 per cent, while at the same time improving the services provided on the site and the quality of the product promised to the commissioning clients.

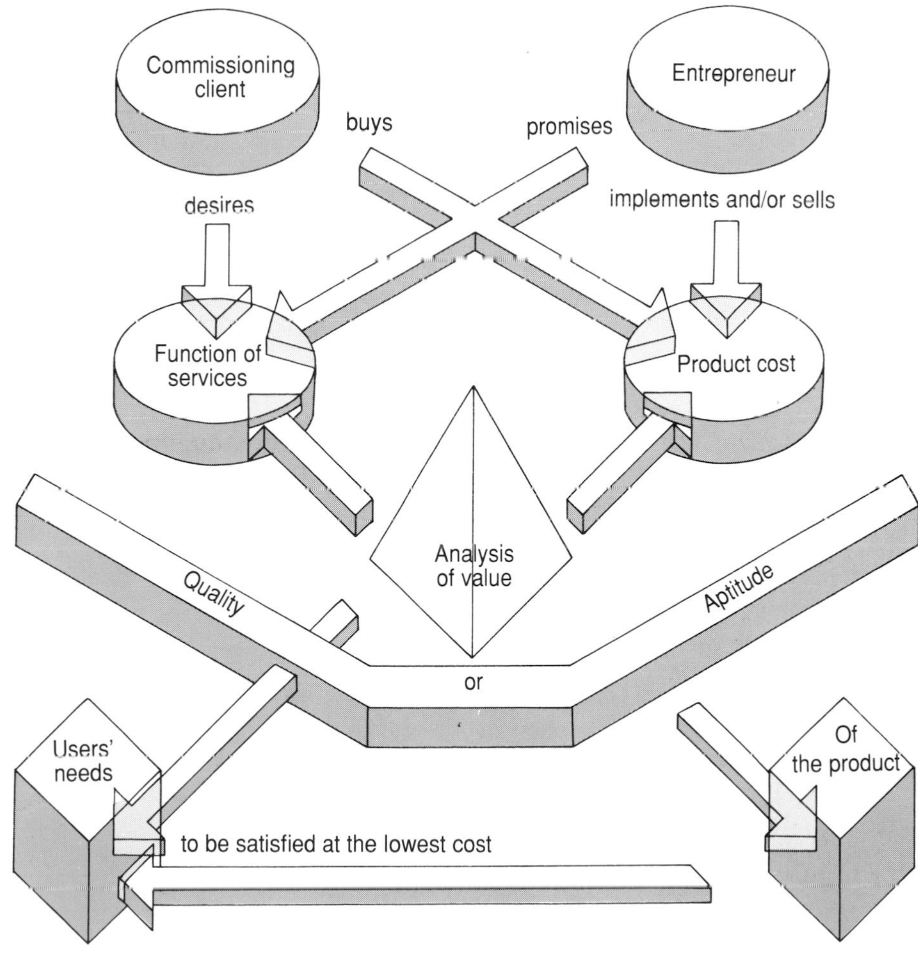

Fig. 4.4

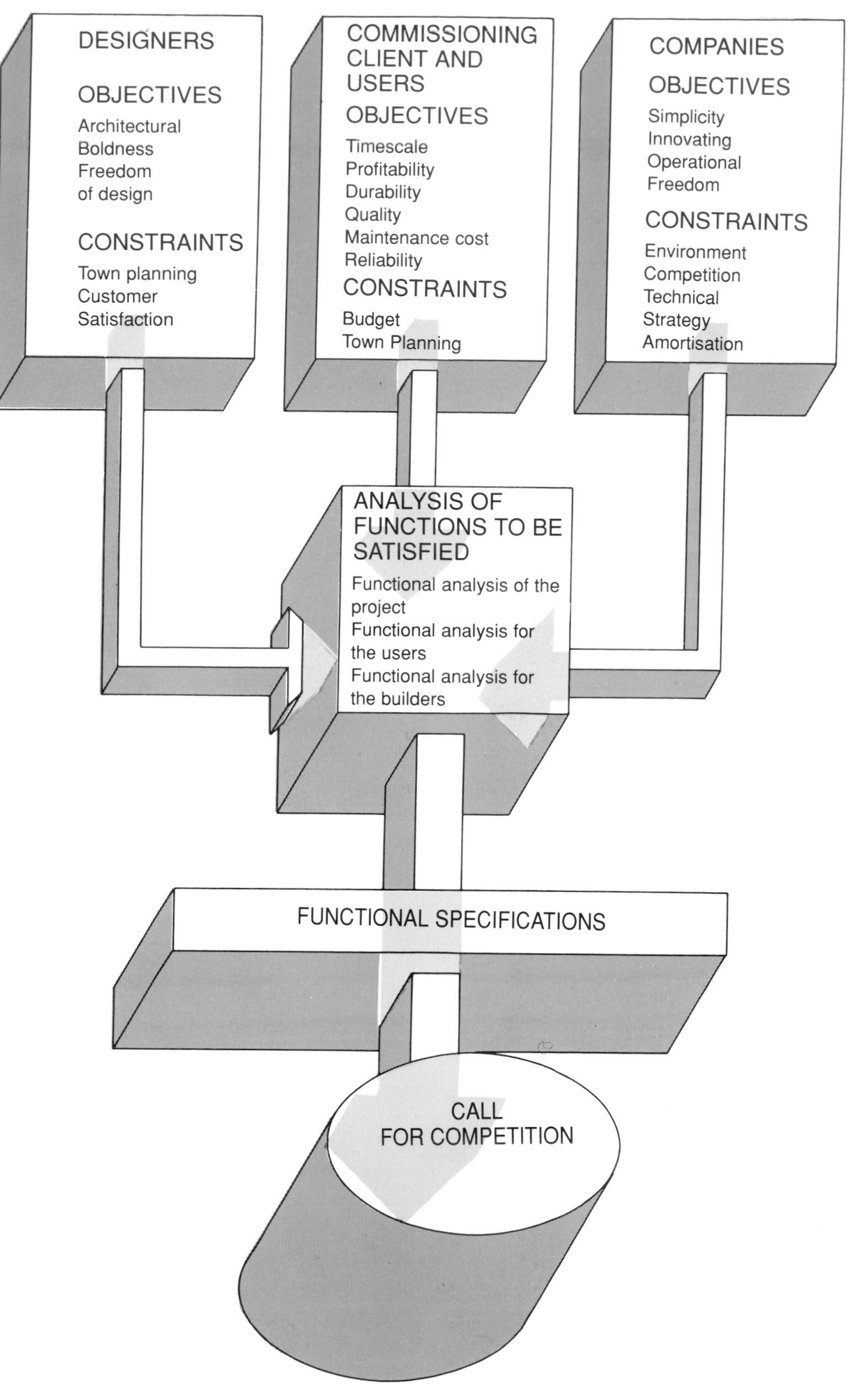

Fig. 4.5 Project functions tree

This optimisation is the result of a functional analysis (Fig. 4.5) which is divided into:

– 'objective' functions;
– 'constraint' functions

which are brought together in the form of a functions tree, which formalises the objectives which will later be written into a **list of functional specifications**.

4.3.1 NEW RULES MUST BE FORMULATED

The specifications should devote less attention to describing the procedures and resources to be employed. They will specify more clearly the objectives to be attained, so as to respond as exactly as possible to the needs of the users.

But this presupposes a training for future architects, company managers, engineers, site managers, workers, etc., leading to a shared and enhanced **project culture**.

4.3.2 FUNCTIONAL SPECIFICATIONS (16)

These are shown in Fig. 4.6.

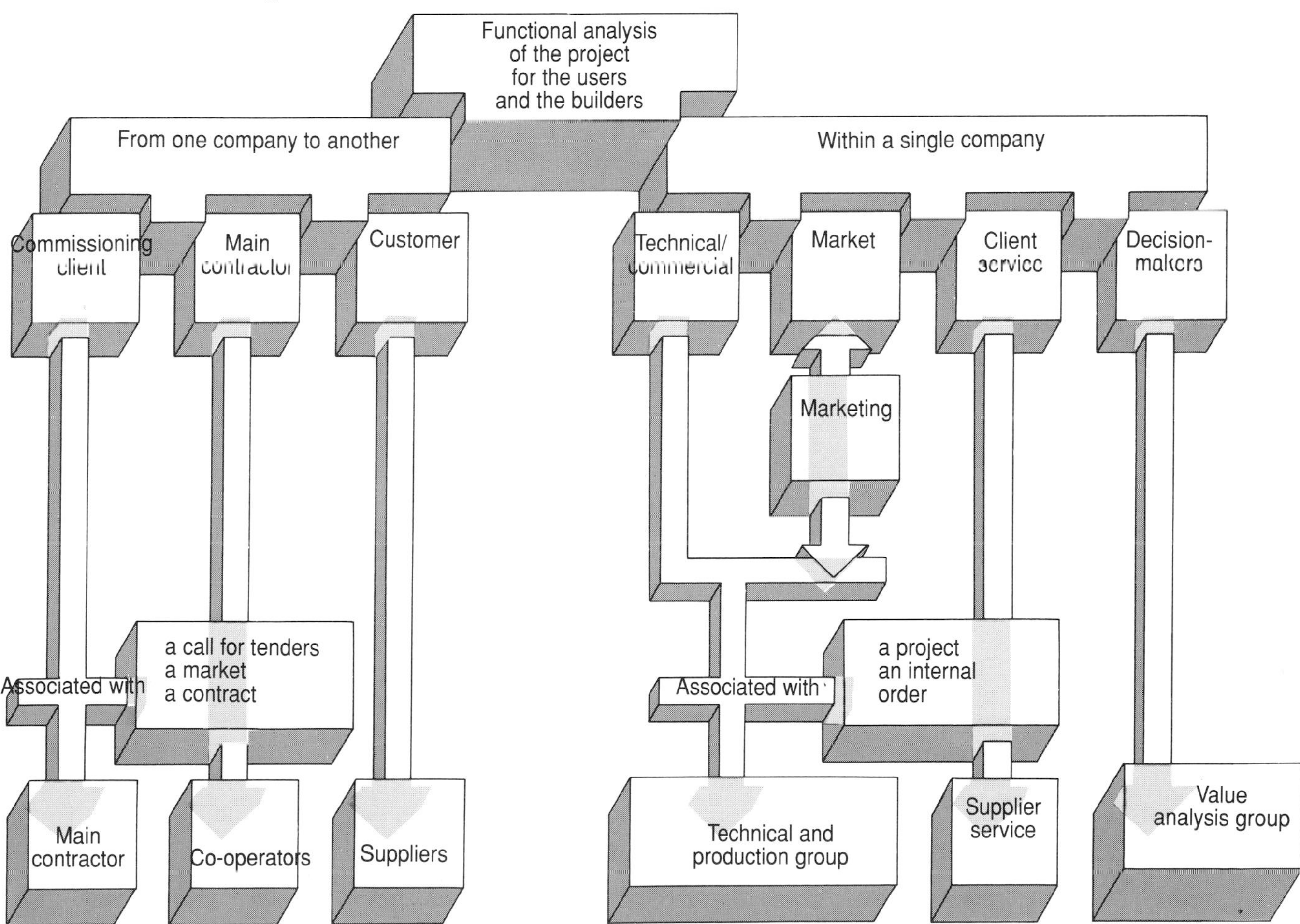

Fig. 4.6

4.4 CONCLUSIONS

The important thing for decision-makers is to know that there are methods and tools which make it possible to manage the process effectively. These tools and methods are based on an improved expression and transcription of needs.

The analysis of value, quality and ergonomics are three approaches:

- the first of which catalogues the functions by specifying the 'global' requirements;
- the second of which specifies the 'correct' quality requirements;
- the third of which measures the 'cost of human work', in order subsequently to improve the comfort and productivity of the work.

These three disciplines all bring together, at different stages, communication, multidisciplinary principles, areas of competence and judgements from the different partners, extending over the entire process.

CHAPTER 5
THE NEW STAGES IN AN ARCHITECTURAL PROJECT

The concept of buildability means, at the design stage, the degree to which measures are considered which facilitate the building of the project, while responding to the functions required from the completed job (Fig. 5.1) (17). As we saw in Chapter 4, some of these parameters have to be taken into account right from the orientation phase. Thus, from this stage onward, there may be a clash between the energy requirements of the project and the short-term financial return or a clash between aesthetic considerations and maintenance needs . . .

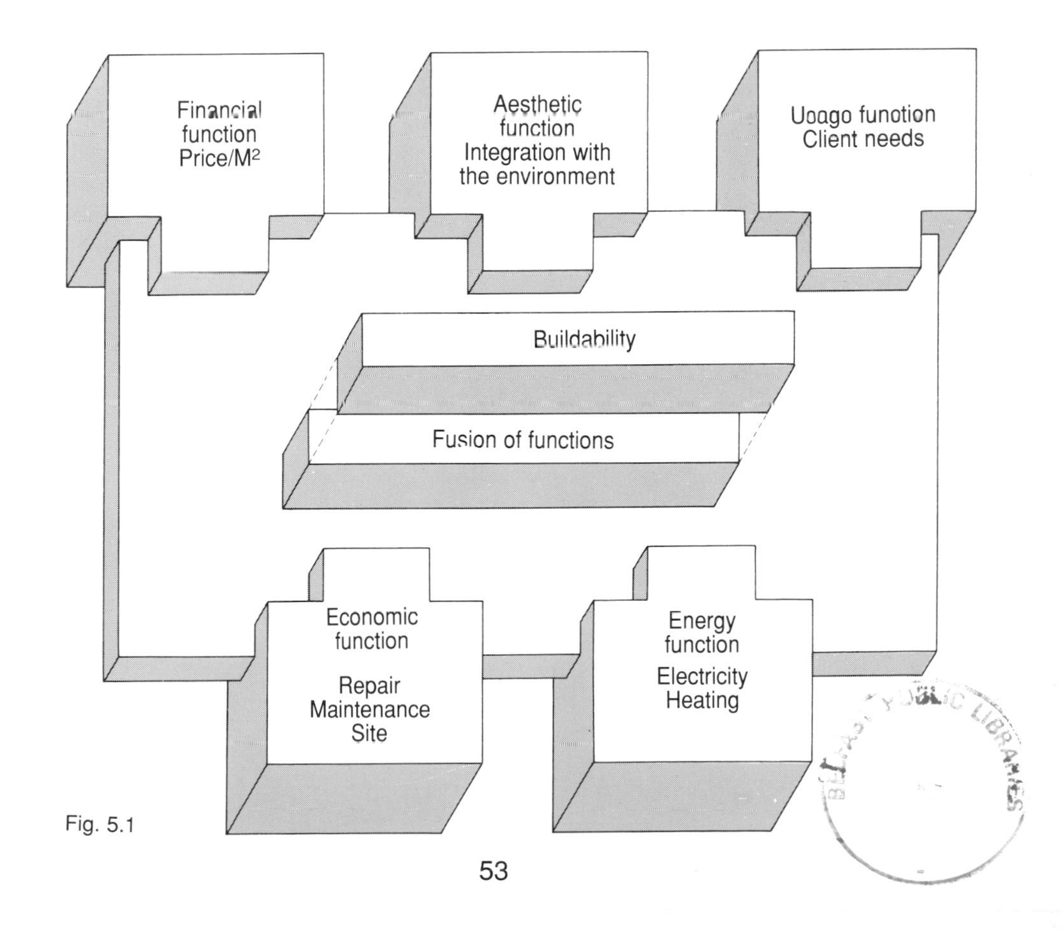

Fig. 5.1

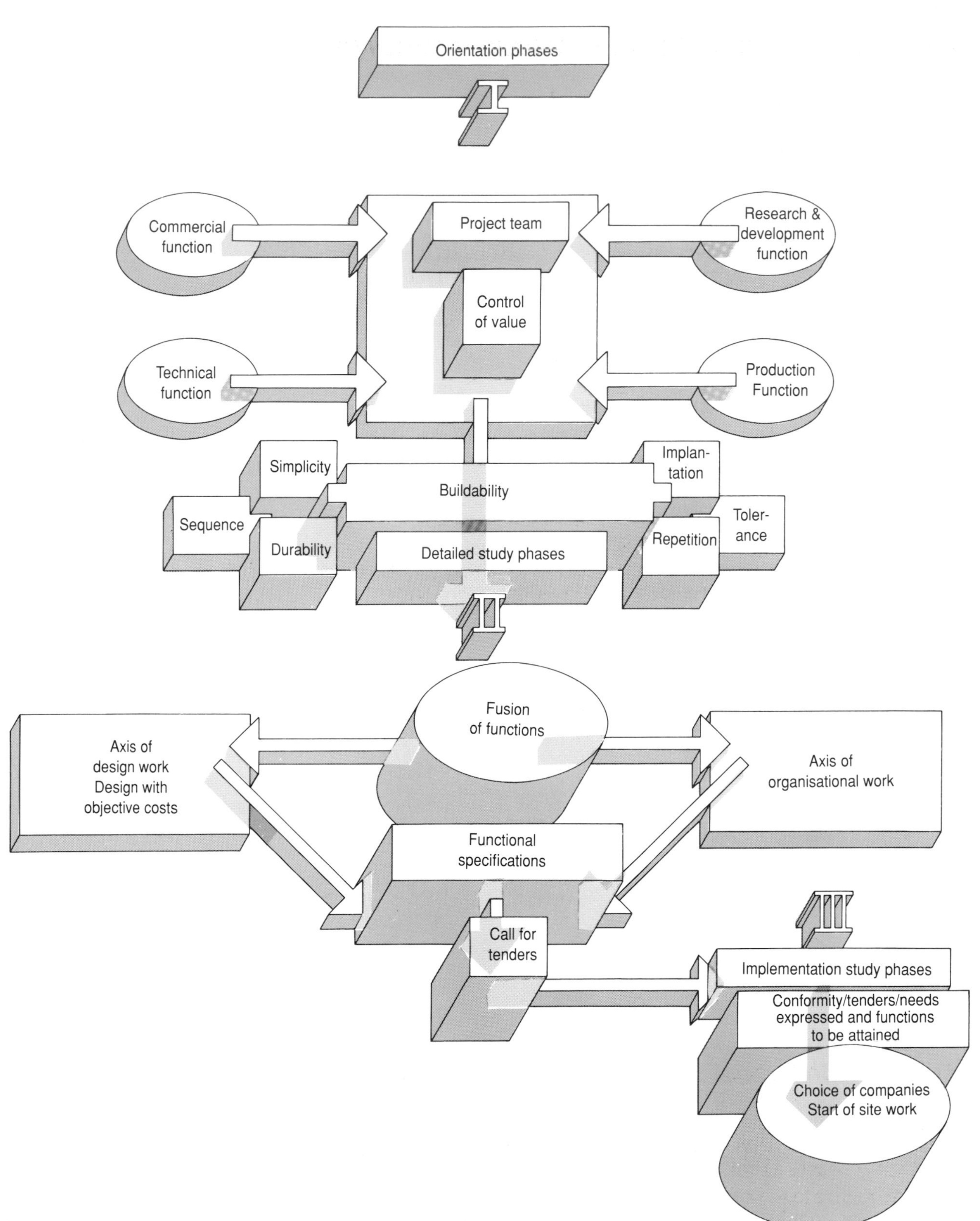
Orientation phases
I
Commercial function
Project team
Research & development function
Control of value
Technical function
Production Function
Simplicity
Implan-tation
Buildability
Sequence
Toler-ance
Durability
Repetition
Detailed study phases
II
Fusion of functions
Axis of design work Design with objective costs
Axis of organisational work
Functional specifications
III
Call for tenders
Implementation study phases
Conformity/tenders/needs expressed and functions to be attained
Choice of companies Start of site work

Fig. 5.2

5.1 A SYSTEMATIC PROJECT MANAGEMENT APPROACH

The life of the project is divided into two parts (Fig. 5.2).

5.1.1 THE PRE-CONTRACT PERIOD

This represents an important phase on account of its cost and duration.

5.1.1.1 Orientation phase

This will be devoted to the analysis of objectives and constraints by a project team which will determine, on the basis of the specification of the needs of each interested party, whether the project matches up to the concept of buildability. It is at this stage that the fusion of functions is carried out (14,16).

5.1.1.2 Detailed study phase before call for tenders

The project team will draw up the functional list of specifications, reconciling the design and organisational constraints. This period comes to an end with the call for tenders.

5.1.2 THE CONTRACT PERIOD

5.1.2.1 Implementation study phase within the companies

The client will order the project on the basis of tenders received, responding to the functional specification list which may perhaps have been amended by those submitting tenders. The project team will check that the tenders match the specifications.

5.1.2.2 Choice of companies – beginning of site work

This is the stage at which the contract phase, i.e. the initial phase of implementation, begins. It leads up to the establishment of a special structure, with the aim of making a complex group of companies function together.

5.2 AN IDEALISATION OF CONFORMITY APPROACH

Securing improvements, cutting costs and reducing occupational accidents involve the mobilisation of a whole team during the orientation phase and the pre-contract study phase, as well as while the site work is going on.

A new site corresponds to a cycle which gets under way, expressing a continual idealisation of the conformity of services and products to the needs which have been expressed (14).

The cycle of a building site (Fig. 5.3) does not imply any order of priority for the choice of functions when the programme gets under way.

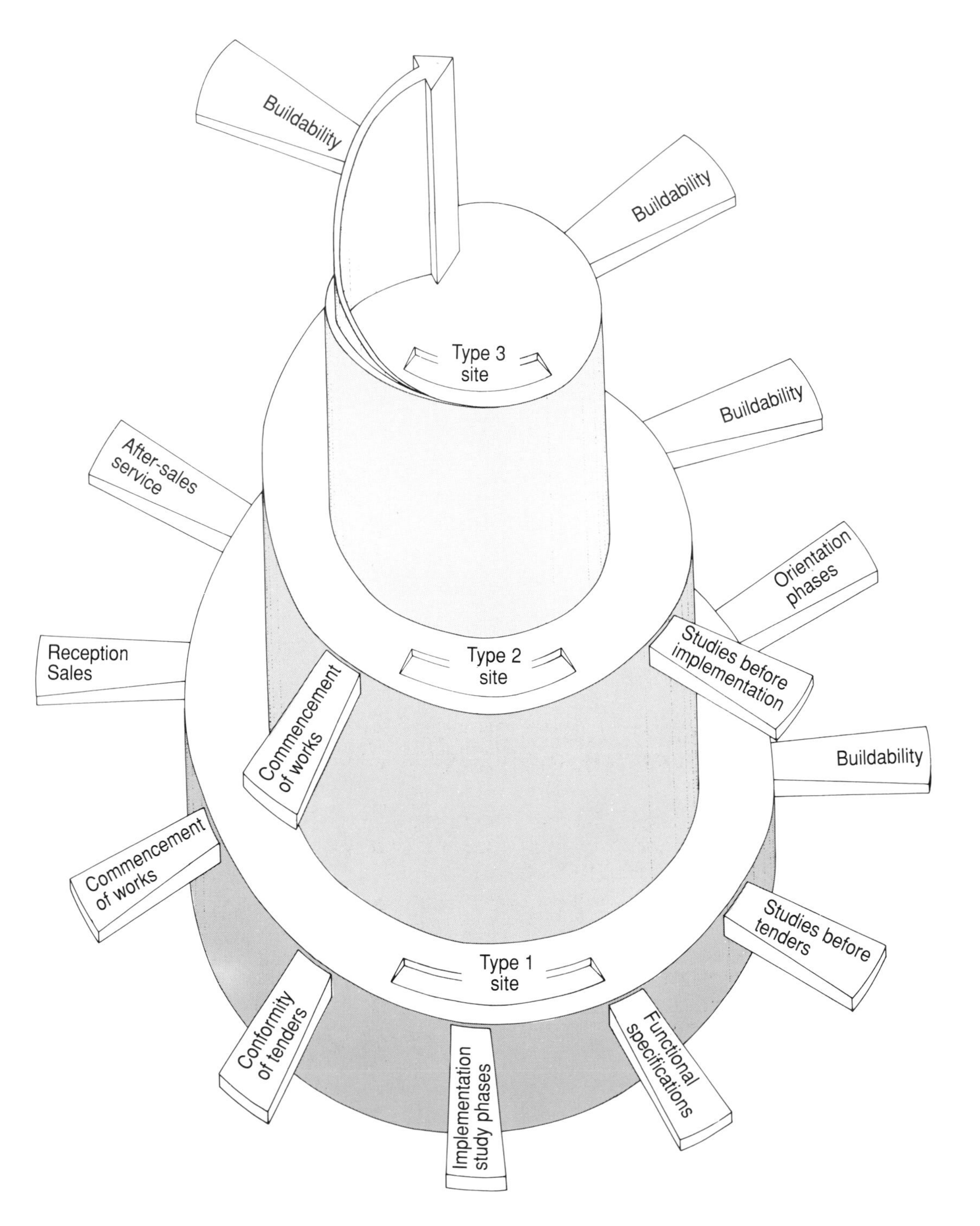

Fig. 5.3

5.3 COST OF PREVENTION OF SHORTCOMINGS

The integration of the concepts of ergonomics, value analysis and total integrated quality management over the whole construction process, undoubtedly represent an extra cost (longer preparation time), but this extra cost will quickly be paid off and will become a source of productivity (Fig. 5.4) (14).

Fig. 5.4 (a) *Improving* used to mean eliminating defects after production, thereby increasing expenditure on repairs and reducing the number of hours that could be devoted to other sites; (b) it now means taking preventive action as far as possible in advance, in order to reduce potential errors.

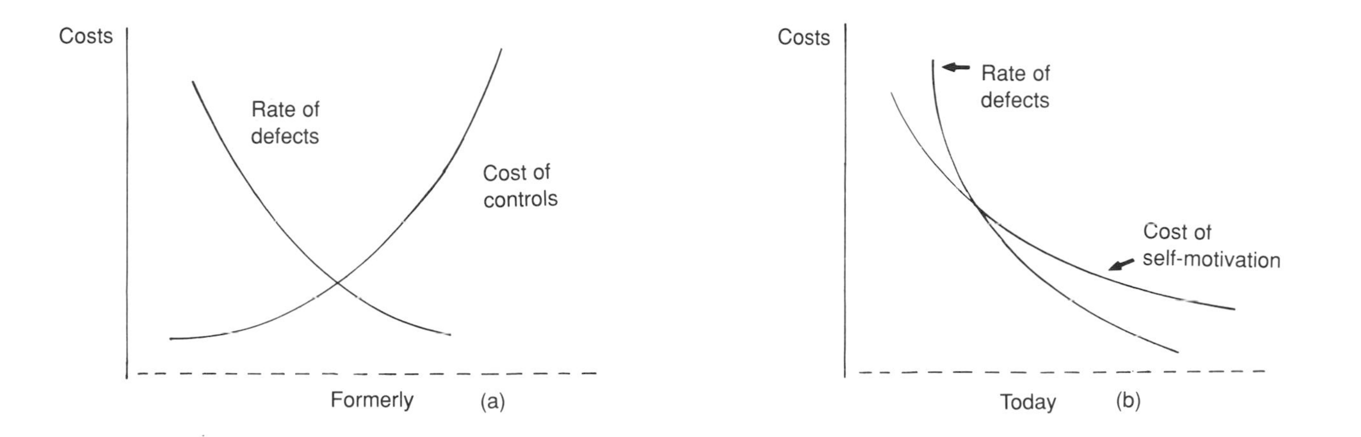

5.4 ANTICIPATING THE RISKS OF SOCIAL AND ECONOMIC COSTS AT THE RIGHT TIME

It is important to eliminate these risks as soon as they can be identified. A risk which has not been identified at the right moment gives rise to a higher cost than the prevention cost would have been.

For each of the stages which we have defined on page 55, it is possible to evaluate the risk of potential errors. The decision to move to the following stage can only be taken when this risk has been assessed at an acceptable level. The repeated stages (specification of needs, fusion of functions, evaluation of risks) will be gone through during the orientation phases in the first period, and will be related to the following systems and subsets:

- participants;
- operations;
- products;
- choice of companies;
- composition of teams.

This approach is equally applicable during the orientation period and during the study phases, as well as when equipment (shuttering, ventilation grilles, furnishings, etc.) is being chosen (Figs 5.5 and 5.6).

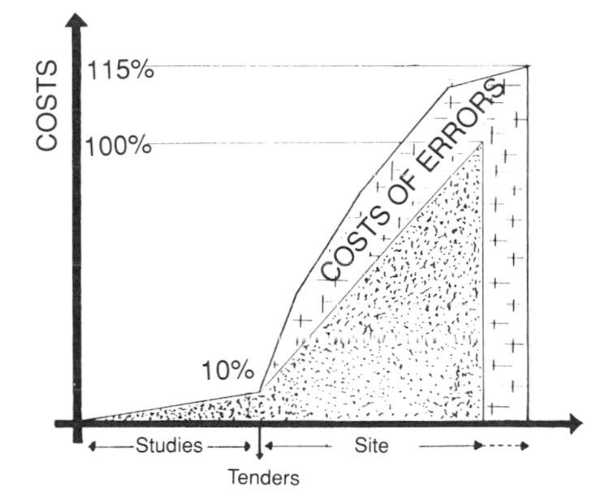

Fig. 5.5 Traditionally, the time given over to studying the project is much shorter than the time devoted to site work. Now, the financial performance of the operation is partly dependent on the site period during which 90% of the budget is unproductive, the remaining 10% being earmarked for studies. We have seen that the cumulative rate of social and economic poor-quality could reach between 10 and 18% of the total cost of the operation.

Fig. 5.6 Controlling identifiable risks (16).

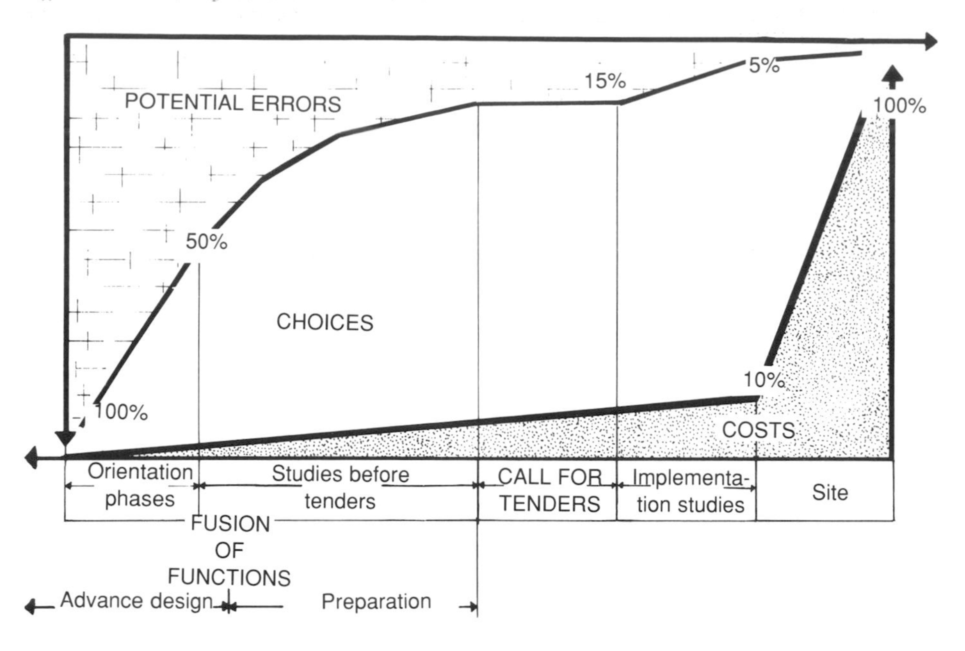

5.5 STRATEGY-DEPENDENT APPLICATION OF THIS APPROACH

Three main tendencies emerge, and can coexist (2).

5.5.1 THE PATH OF INDUSTRIALISATION

This path supposes that the final product is to be clearly negotiated and perfectly defined in advance together with the client. The nature and choice of components determine a sequential type of organisation which:

- relies on close co-operation between companies and industrialists;
- creates new interfaces between the companies carrying out the project;
- goes together with new planning and co-ordination methods;
- relies on new multi-function skills in the management and implementation areas.

5.5.2 THE COMPANY ENGINEERING PATH

This approach, described as 'performance-based', includes simultaneously:

- the establishment of a 'dialogue', well in advance, between architects and companies;
- design and implementation engineering involving:
  - from the design phase onward, a consideration of the elements of organisation, cost and expertise of companies charged with implementing the project;
  - the ability of those implementing the project to communicate the detailed knowledge of the different techniques and conditions for their utilisation, right from the design stage;

- a new type of relationship between the building trades, involved;
- a process of participation in the advance studies between large companies and small and medium-sized enterprises.

5.5.3 THE WORKFORCE APPROACH

This strategy, a minority one in the building industry, has to do with a desire to enlarge the field of intervention for the teams:

- by giving back to the building site a certain degree of control and autonomy, starting from the phase of preparation and organisation of the works;
- by developing the motivation and involvement of the workers.

This approach is most obviously applicable in the case of renovation works, maintenance works and small sites.

CHAPTER 6
CONCLUSIONS

This new approach produces three pathways to innovation:

- a cluster of client–supplier relationships;
- a cluster of processes and products;
- an integrated organisation where consultation replaces part of the decision-making process.

6.1 A CLUSTER OF CLIENT–SUPPLIER RELATIONSHIPS

The consideration of the site as a cluster of client–supplier relationships involves two concepts which offer considerable possibilities for the promotion of product quality and ergonomics in working conditions.

6.1.1 THE CLIENT–SUPPLIER RELATIONSHIP

The quality of working conditions and product quality, considered as matching the needs emerging from a client–supplier relationship, is a system value which implies exchange between individuals or groups. It develops communication, feedback, and connections between neighbouring elements. It is thus a source of immediate and permanent refinement and progress.

On the other hand, when quality is considered as merely matching a set of specifications, it is a value which brings people face to face with a document over which they have had no influence, or very little influence; it confines the individual within a field of action where his or her opportunities to secure improvements are limited.

6.1.2 THE FORMAL CONCEPT OF THE CLIENT

Up to the present, the concept of a client meant customers outside the company, or outside the professional survey office. From now on, the concept of a client is applicable to all those charged with implementation, all those working on design, who are both clients and suppliers, and thereby responsible for the quality of the product and working conditions in relation to the recipient of their work output.

Experience gained in this area shows that this vision, which is conceptually very simple, is nevertheless a very fertile one.

6.2 A CLUSTER OF PROCESSES AND PRODUCTS

The consideration of the site as a cluster of functional definitions, generating new services and products, is the second innovative element.

A more thoroughgoing analysis of methods leads to the site being broken down into a set of processes and actors generating internal and external products.

The methods of improvement are applied then to each of these factors, which involves the introduction of new conceptual and organisational approaches in areas of activity which were previously relatively unexplored: the project functions for these areas are expressed in terms of objectives rather than resources.

Objectives for one area will sometimes be constraints for another, but there are shared approaches which make it possible to implement a coherent action policy to reduce, in a systematic fashion, potential errors, sources of accidents, of poor quality and poor productivity, at all stages of the project.

6.3 AN INTEGRATED ORGANISATION

Lastly, the integrated organisation of the site from the orientation phases onward seems to us to be the third innovation.

It is this integrated organisation which ensures the optimisation of the creative contribution of all actors, and consequently a better consideration, by creators and implementors, of the problems arising in the areas of safety and ergonomics, product quality and productivity and costs. This is done so that the project may be easily implemented, a natural and essential concern of all those involved.

It is therefore important to secure a good definition of the role, objectives and constraints of each partner:

a) The commissioning client expects a product of quality matching his or her needs, within an agreed price and timescale.

b) The designing architect is in charge of the aesthetic representation, which will be realised with products the functional qualities of which he or she will specify.

This basic choice will determine a method of implementation and a set of resources, the usability of which will depend on the product and the company constraints (the buildability concept).

c) In the companies, the improvement of internal relations, the choice of equipment, and the establishment of hierarchical guidelines are also factors which can bring improvements. The definition of needs in terms of performance rather than resources has a direct effect on innovation. These elements must therefore be handled together from beginning to end, while improving the creative and operational freedom of each partner.

6.4 TOWARDS A PROJECT MANAGEMENT GUIDE

This approach is expressed through a functional interpretation initiative, the main lines of which will be made clear in the project management guide, which is shortly to be produced by the European Foundation for the Improvement of Living and Working Conditions.

BIBLIOGRAPHY

The publications listed below are referred to in the text in abbreviated form, by the bracketed number in the left-hand column

(1) BIRCHALL D W, *Working Conditions in the Construction Industry*, Henley, the Management College, 1987.

(2) CAMPAGNAC E et CARO C, *Les conditions de travail dans l'industrie de la construction*, CERTES, Paris, 1987.

(3) DIEPEVEEN J et BROUWERS A, *Les conditions de travail dans l'industrie de la construction*, Stichting Bouwresearch, Rotterdam, 1987.

(4) LORENT P, *Les conditions de travail dans l'industrie de la construction, productivité, conditions de travail, qualité concertée et totale*, CNAC, Brussels, 1987.

(5) ROLLIER M, *La sicurezza nel settore delle costruzioni come problema di organizzazione e di progettazione*, RSO, 1987.

(6) SPANNHAKE B, *Les carences de la sécurité du travail dans l'industrie de la construction et les frais qui en découlent*, Dortmund, 1987.

(7) ALBIZZATI M, *Quel sera l'impact du marché européen sur le bâtiment?* Le moniteur no 8 du 20 février 1987 (France).

(8) POPPY W, *Analyse des risques professionnels dans l'industrie de la construction pour la Commission des Communautés Européennes en collaboration avec la chaire «Technique et machines de construction»* (Dortmund) et la Bau-Berufsgenossenschaft (Wuppertal).

(9) GORISSE F, *Coûts, Qualité, Saisie des coûts de non-qualité*. Qualiform, Vanves, Paris.

(10) LETOUBLON M, *Politique et stratégie des entreprises du bâtiment et travaux publics en matière d'accidents de travail*, Lyon, L'Hermès, 1979.

(11) RAGU M, PORTIER M, PELTIER P, HO M T, *Analyse des causes d'accidents dus aux chutes de hauteur dans l'industrie du bâtiment et des travaux publics. Indications en vue de leur prévention*, INRS Report no. 560/RE 1981, Paris.

(12) CNUDDE M, *Kwaliteit spaart 50 milliard* BF, CSTC, Brussels.

(13) SCHILS J P, *L'ergonomie appliquée au secteur de la construction: enjeu de la sécurité intégrée*, Mémoire, Université Catholique de Louvain, 1987.

GEENS Ph, *Enjeu qualité*, 1986, Mémoire de formation complémentaire en sécurité et hygiène (CNAC).

(14) STORA G, MONTAIGNE J, *La Qualité totale dans l'entreprise*. Management 2000, Paris, Les éditions d'organisation, 1986.

(15) LORENT P, *Ergonomie et Construction*, Publication du Comité National d'Action pour la Sécurité et l'Hygiène dans la Construction, Brussels, 1984.

(16) *Maîtrise de la valeur*, 3rd and 4th conferences, 28 and 29 March 1984 and 16 and 17 April 1986, published by la AFAV, Paris

(17) *Buildability: an assessment* (Special Publication 26), Construction Industry Research and Information Association, London.

(18) *Construction Safety*, London Building Advisory Service, issues from November 1969 onwards.

(19) HSE, Blackspot Construction: A study of five years fatal accidents in the building and civil engineering industries, London, HMSO, 1988.

(20) HSE, *Essentials of Health and Safety at Work*, London, Health and Safety Executive, 1988.